Studies in Military and Strategic History

General Editor: **William Philpott**, Professor of the War Studies, King's College London

Published titles include:

Martin Alexander and William Philpott (*editors*)
ANGLO–FRENCH DEFENCE RELATIONS BETWEEN THE WARS

Christopher M. Bell
THE ROYAL NAVY, SEAPOWER AND STRATEGY BETWEEN THE WARS

Peter Bell
CHAMBERLAIN, GERMANY AND JAPAN, 1933–34

Antony Best
BRITISH INTELLIGENCE AND THE JAPANESE CHALLENGE IN ASIA, 1914–41

Antoine Capet (*editor*)
BRITAIN, FRANCE AND THE ENTENTE CORDIALE SINCE 1904

Philippe Chassaigne and Michael Dockrill (*editors*)
ANGLO-FRENCH RELATIONS, 1898–1998
From Fashoda to Jospin

Michael Dockrill
BRITISH ESTABLISHMENT PERSPECTIVES ON FRANCE, 1936–40

Michael Dockrill and John Fisher
THE PARIS PEACE CONFERENCE, 1919
Peace without Victory?

John P. S. Gearson
HAROLD MACMILLAN AND THE BERLIN WALL CRISIS, 1958–62

Brad William Gladman
INTELLIGENCE AND ANGLO-AMERICAN AIR SUPPORT IN WORLD WAR TWO
The Western Desert amd Tunisia, 1940–43

Raffi Gregorian
THE BRITISH ARMY, THE GURKHAS AND COLD WAR STRATEGY IN
THE FAR EAST, 1947–1954

Stephen Hartley
THE IRISH QUESTION AS A PROBLEM IN BRITISH FOREIGN POLICY, 1914–18

Ashley Jackson
WAR AND EMPIRE IN MAURITIUS AND THE INDIAN OCEAN

Jonathan Krause
THE GREATER WAR
Other Combatants and Other Fronts, 1914–1918

James Levy
THE ROYAL NAVY'S HOME FLEET IN WORLD WAR II

Stewart Lone
JAPAN'S FIRST MODERN WAR
Army and Society in the Conflict with China, 1894–95

Chris Mann
BRITISH POLICY AND STRATEGY TOWARDS NORWAY, 1941–45

Thomas R. Mockaitis
BRITISH COUNTERINSURGENCY, 1919–60

Bob Moore and Kent Fedorowich
THE BRITISH EMPIRE AND ITS ITALIAN PRISONERS OF WAR, 1940–47

T. R. Moreman
THE ARMY IN INDIA AND THE DEVELOPMENT OF FRONTIER WARFARE, 1849–1947

Kendrick Oliver
KENNEDY, MACMILLAN AND THE NUCLEAR TEST-BAN DEBATE, 1961–63

Paul Orders
BRITAIN, AUSTRALIA, NEW ZEALAND AND THE CHALLENGE OF THE UNITED STATES, 1934–46
A Study in International History

Elspeth Y. O'Riordan
BRITAIN AND THE RUHR CRISIS

G. D. Sheffield
LEADERSHIP IN THE TRENCHES
Officer–Man Relations, Morale and Discipline in the British Army in the Era of the First World War

Adrian Smith
MICK MANNOCK, FIGHTER PILOT
Myth, Life and Politics

Melvin Charles Smith
AWARDED IN VALOUR
A History of the Victoria Cross and the Evolution of British Heroism

Nicholas Tamkin
BRITAIN, TURKEY AND THE SOVIET UNION, 1940–45
Strategy, Diplomacy and Intelligence in the Eastern Mediterranean

Martin Thomas
THE FRENCH NORTH AFRICAN CRISIS
Colonial Breakdown and Anglo-French Relations, 1945–62

Simon Trew
BRITAIN, MIHAILOVIC AND THE CHETNIKS, 1941–42

Kristian Coates Ulrichsen
THE LOGISTICS AND POLITICS OF THE BRITISH CAMPAIGNS IN THE MIDDLE EAST, 1914–22

Steven Weiss
ALLIES IN CONFLICT
Anglo-American Strategic Negotiations, 1938–44

Studies in Military and Strategic History
Series Standing Order ISBN 978–0–333–71046–3 Hardback
978–0–333–80349–3 Paperback
(*outside North America only*)

You can receive future titles in this series as they are published by placing a standing order. Please contact your bookseller or, in case of difficulty, write to us at the address below with your name and address, the title of the series and the ISBN quoted above.

Customer Services Department, Macmillan Distribution Ltd, Houndmills, Basingstoke, Hampshire RG21 6XS, England

Mick Mannock, Fighter Pilot

Myth, Life and Politics

Adrian Smith
University of Southampton, UK

in association with
King's College London

© Adrian Smith 2001, 2015

All rights reserved. No reproduction, copy or transmission of this publication may be made without written permission.

No portion of this publication may be reproduced, copied or transmitted save with written permission or in accordance with the provisions of the Copyright, Designs and Patents Act 1988, or under the terms of any licence permitting limited copying issued by the Copyright Licensing Agency, Saffron House, 6–10 Kirby Street, London EC1N 8TS.

Any person who does any unauthorized act in relation to this publication may be liable to criminal prosecution and civil claims for damages.

The author has asserted his right to be identified as the author of this work in accordance with the Copyright, Designs and Patents Act 1988.

First published 2001
Published in paperback 2015 by
PALGRAVE MACMILLAN

Palgrave Macmillan in the UK is an imprint of Macmillan Publishers Limited, registered in England, company number 785998, of Houndmills, Basingstoke, Hampshire, RG21 6XS.

Palgrave Macmillan in the US is a division of St Martin's Press LLC, 175 Fifth Avenue, New York, NY 10010.

Palgrave is the global academic imprint of the above companies and has companies and representatives throughout the world.

Palgrave® and Macmillan® are registered trademarks in the United States, the United Kingdom, Europe and other countries.

ISBN 978–0–333–77898–2 hardback
ISBN 978–1–137–50982–6 paperback

This book is printed on paper suitable for recycling and made from fully managed and sustained forest sources. Logging, pulping and manufacturing processes are expected to conform to the environmental regulations of the country of origin.

A catalogue record for this book is available from the British Library.

A catalog record for this book is available from the Library of Congress.

Typeset by MPS Limited, Chennai, India.

To Jeanne Ogilvie and Anne Tooke

Contents

Map: Mick Mannock's Western Front	ix
Preface to the Paperback Edition	x
1 Introduction	1
2 A Prewar Education, 1888–1914	11
From Cork to Canterbury	11
Raising the red flag in the heart of the Midlands	25
3 Preparing for War, 1914–17	39
Confinement in Constantinople	39
Agitation in uniform	48
Planes before politics – learning to fly	61
4 40 Squadron, 1917–18	68
An uncomfortable apprenticeship	68
Flight commander	85
5 74 and 85 Squadrons, 1918	100
Sojourn in Blighty, January–March 1918	100
Veteran flight commander	107
Life and death with 85 Squadron, July 1918	122
'Untroubling and untroubled where I lie/The grass below, above the vaulted sky'	128
6 Conclusion	132
Remembering Mick	132
Socialist air ace	154
Notes	163
Bibliography	198
Index	205

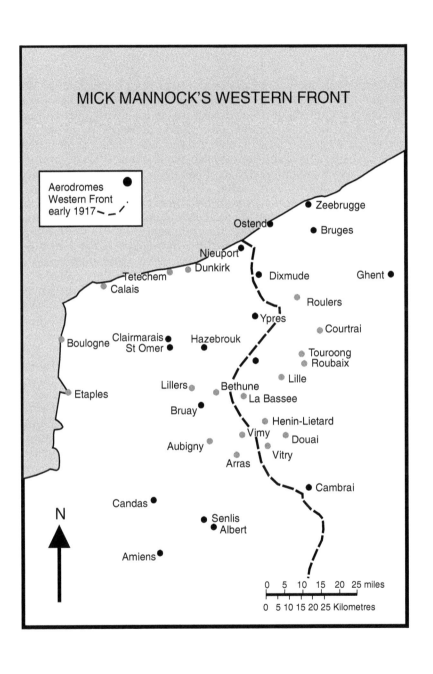

Preface to the Paperback Edition

War memorials come in all shapes and sizes, from rugby grounds like Harlequins' home, The Stoop, through to infirmary beds, such as the one in Kent and Canterbury Hospital dedicated to ace fighter pilot Edward Mannock. Across eight decades, on a Sunday afternoon in mid-July, a succession of bemused patients observed a minute's silence at the behest of their visitors. The ward visit rounded off an annual ritual in memory of Canterbury's most decorated war hero. Sadly, the hospital pilgrimage and its preceding commemorative service are no more. My inaugural attendance at the ceremony was in the summer of 1998, by which time Mannock had been dead eighty years. The anniversary of his death was eclipsed by elaborate recollection of more momentous events in the final year of the First World War, not least the signing of the armistice. Nevertheless, as they had done every summer for over thirty years, a tiny band of relatives, civic dignitaries, and RAF veterans gathered in front of a memorial tablet in the nave of Canterbury Cathedral to remember a man admired by his comrades as the ace of aces.

'Mick' Mannock served on the Western Front from April 1917 until his death on 26 July 1918, during which time he rose from 2nd lieutenant to (acting) major and squadron commander, and was awarded the DSO and two bars, and the MC and bar. Released from Turkish internment he had re-enlisted in his Territorial Army ambulance unit in July 1915, but then sought active service as a sergeant in the Royal Engineers. Newly commissioned, in August 1916 Mannock had transferred to the Royal Flying Corps, which on 1 April 1918 would form one half of the newly formed RAF. Officially credited with 50 victories, Major Mannock's unofficial score was later calculated as 73; and on 18 July 1919 he was posthumously awarded the Victoria Cross. On that warm Sunday afternoon in east Kent in 1998 I applauded a worthy warrior of the skies. At the same time I asked myself if it wasn't time that awareness of this half-forgotten hero should spread beyond the specialist *and* the enthusiast. Perhaps the life – and equally the myth – of 'Mick' Mannock warranted closer attention, if only because both promoted so many questions about memory, biography, patriotism, commemoration, heroism, *and* the complex relationship between the public and the private, the political and the personal. A further incentive was that previous lives of my man were on the whole pretty poor – I quickly came to the conclusion that more interesting than the books themselves was why and when they were written. So, here was someone who deserved a properly researched biography, locating its subject's frontline experience within the wider story of wartime aviation's astonishing transformation.

Mick Mannock, Fighter Pilot was my first attempt at writing a full-length biography, and as such an invaluable apprenticeship. Not that I feel the book suffered in any way from my learning on the job. I was after all an experienced historian in his forties who had written and published widely, and had a 'big book' under his belt; this was a history of the early *New Statesman*, written long after completion of my PhD, and in consequence far superior to any immediate translation of thesis into monograph. A biography of Edward Mannock was a natural progression from researching and teaching about the Edwardian labour movement, and from my parallel preoccupation with global conflict and civil–military relations. A keen interest in matters military was an inevitable consequence of having taught at Sandhurst. I only spent two years at the Royal Military Academy but it was a formative experience, forcing me to refocus my research, to challenge lazy assumptions about the complex relationship between the armed forces and wider society in contemporary Britain, and to embrace a fresh intellectual agenda. Under the watchful eye of Matthew Midlane, a future director of academic studies, for six days of the week I feebly endeavoured to match Alex Danchev's intellectual firepower; and every Sabbath interrogated John Keegan on combat motivation and unit morale in the bizarre setting of the Camberley laundrette.

The story of *Mick Mannock, Fighter Pilot* ends in July 1918, except that it doesn't in so far as the final chapter focuses upon memorialization, mythologizing, and the posthumous representation of 'the ace with one eye'. Yet a shrewd reader of these late reflections upon our hero's afterlife would recognize the book as very much a turn-of-the-century text. The opening chapters, which gradually reveal the roots of the book, suggest an author susceptible to the nostalgia of middle age but at the same time intent on injecting idealism into the hard-nosed policy priorities of Tony Blair's Britain. The conclusion proved prescient in anticipating the degree to which the Labour leader and his closest advisers enthusiastically embraced military action. Their motives may in the first instance have been honourable, witness 'humanitarian intervention' in Sierra Leone, but they lamentably failed to comprehend the consequences of their actions. This lack of insight and absence of empathy – an inability, or for Blair's harshest critics a reluctance, to understand what was actually taking place on the ground – arises out of career politicians and their special advisers (and indeed most senior civil servants) having no direct military experience; nor, seven decades after the Second World War, and five since the end of National Service, having frequent contact with family or friends who at some point in their lives served in the armed forces.

Military advisers inside Whitehall can spell out the harsh reality of sending British personnel into a combat zone, but successive chiefs of the defence staff have received stiff reminders from prime ministers of both major parties where executive power rests in the British model of command and control. Successive cabinets since 1997 have lacked a Willie Whitelaw or a Peter Carrington, uniquely qualified to have a quiet word in the PM's ear and spell out the full implications of any particular operational decision. Cabinet caution in April 1982 arose out of senior ministers recognizing the gravity of their ordering a task force to restore British sovereignty over the Falklands: in an extraordinarily risky operation there was a strong possibility of high casualties, incurred while fighting in extreme conditions. Notwithstanding Mrs Thatcher's jubilation over the recapture of South Georgia, ministers were notably downbeat until after the Argentinian surrender in Port Stanley. Contrast such behaviour with the unreal predictions of government spokesmen and their political masters, not least defence secretary John Reid, when three thousand troops were despatched to Helmand province in January 2006. It's scarcely surprising that Conservative MP Rory Stewart (Balliol, Black Watch, Foreign Office, intrepid traveller) speaks with authority on Iraq and Afghanistan but remains a backbencher.

Ironically, the less those in government know about what really takes place out in the field the more visible the military become in everyday civilian life. In this respect the British since 9/11 have aped the United States, but not so comprehensively as to validate a charge of becoming more militaristic. We must of course be wary of sweeping assertions. Mannock's deep hostility towards his enemy was rooted in the erroneous belief that the whole of Germany had succumbed to Prussian militarism. This was ironic in that here was a committed socialist seemingly oblivious to Europe's largest social democratic party. Nevertheless, such an indictment is scarcely surprising given the harsh consequences for Mannock of the German delegation's power and influence once Turkey entered the war. He returned from Constantinople in the spring of 1915 a bitter man, indifferent to any suggestion that the Kaiser's empire contained within it a range of opinions and beliefs, not least an ideology akin to his own which had permeated through to every level of Germany's urban working class. We must be honest and recognize that, for the rest of Edward Mannock's short life, international proletarian solidarity was an ideal he treated with scorn and derision. No doubt his view of the Germans would have mellowed had he been alive between the wars, but the Nazi seizure of power, and the apparent popularity of Hitler's regime, would surely have seen old prejudices rise to the surface.

Mannock's Irish Catholic working-class roots were what made the man so appealing as a subject. James McCudden, with perhaps a higher profile across the past hundred years, shared a similar background. Not that they were the only pilots with Irish parents or from Ireland. Publication of my book generated a follow-up article in *History Ireland*, focusing particularly on W.B. Yeats' fallen hero, Robert Gregory. A few years later I was invited to give a talk on Irish aviators of the Great War at Casement Aerodrome, the headquarters of the Irish Air Corps at Baldonnel, near Dublin. There was an obvious irony in my lecturing on Irishmen who considered themselves patriots at a location named after Roger Casement, arrested by the British authorities on his return from Germany in 1916 and condemned to death for treason. That irony was duly acknowledged by my audience, in itself a reflection of how the Irish Republic, nearly a century after the Easter Rising, has the maturity and self-confidence to salute the 147,000 men and women who from 1914 to 1918 volunteered for service in the British Army or the Royal Navy. Mannock was an archetypal home ruler, supportive of Ireland asserting its cultural and national identity without permanently fracturing the Union. He doesn't appear to have taken offence at being nicknamed 'Paddy' or 'Mick'. Regarding religion he was equally relaxed, his parochial allegiances dictated more by cricket than creed.

The Royal Flying Corps' officers were by no means the homogeneous band of brothers their popular image suggests: the messes of frontline squadrons saw most young men fresh out of public school in awe of their hard-drinking comrades from the colonies. Derek Robinson's scabrous novel *Goshawk Squadron* offers a grim and persuasive portrait of a typical RFC/RAF scout squadron behind the lines on the Western Front. Thus, while class prejudice remained all too apparent, Mannock's proletarian credentials were by no means unique. What made Mannock and McCudden so special was that their working-class origins facilitated an enviable degree of engineering expertise. Most officers with a similar background and a comparable level of mechanical competence, cultivated at specialist institutes such as Finsbury Technical College, were serving in support regiments such as the Royal Engineers or the Royal Corps of Signals. Few, like Mannock, volunteered for transfer to the RFC. Major McCudden was unusual in that he started as a ranker in the Army's youngest corps, and by dint of courage and innate flying ability rose from ground crew to squadron commander.

This combination of natural talent, intelligence, and the technical ability to comprehend why and how an SE5a could be fashioned into a ruthlessly efficient killing machine, was crucial to understanding why Mannock and McCudden were so successful in the skies above the Western Front, and for so long. Layered on to Mannock's distinct qualities as a flier was his politics, thereby rendering him a far more complex personality. When *Mick Mannock, Fighter Pilot* first appeared, the review in *Cross and Cockade*, magazine of the First World War Aviation Historical Society, was predictable. Its author simply failed to see the bigger picture. Of what relevance was our hero's uncompromising socialism, and why should we be interested in his afterlife, other than pinpointing where he died and under which headstone he lies buried? At the other end of the spectrum, the *English Historical Review* was left asking what all the fuss was about, unintentionally revealing how too many historians early this century still saw the study of science and engineering as a specialist niche. Thankfully, over the past decade such intellectual narrow-mindedness has finally faded away, and it's hard now to imagine any study of how the British twice mobilized for 'industrial war' not placing scientific endeavour centre stage. To his great credit John Hayes-Fisher took on board the significance of Mannock's and McCudden's engineering prowess when making his BBC2 documentary *WW1 Aces Falling*. I shall never forget the thrill of seeing the *Timewatch* film on the big screen at the RAF Club when it was premiered in late 2008. Not only did it look terrific, largely thanks to some great aerial shots of replica aircraft in the Shuttleworth Collection, but for the first time it provided a full and accurate portrayal of the man who had captured my imagination ten years before. The producer generously acknowledged the influence my book had in his understanding of what motivated Mannock, and why this difficult man acted in the way that he did – and why he generates generous admiration but only scant affection among those of us who remain fascinated by this ace of aces. Now published for the first time in paperback, on the centenary of Mannock's return from internment in Turkey, the following pages offer a renewed opportunity to share in that same fascination.

Adrian Smith
Lymington, Hampshire

1
Introduction

The borderland that stretches south of the Ypres salient is a grim and miserable place. Here Mick Mannock fought his war. In Flanders fields conflict and industry have come and gone, leaving a harsh legacy. The airfields of 1917–18 have long since been swallowed up in suburban sprawl – scanning the Michelin map for solitary swathes of green simply confirms a landscape transformed by autoroutes and service stations, industrial and retail estates, art deco villas and breezeblock bungalows. Precious little remains of those vast support areas that mushroomed behind the front line once the winter of 1914 signalled stalemate and entrenchment. The complete absence of the original physical environment tests even the keenest imagination.[1] At least Vimy Ridge, Verdun or the killing fields between Albert and Bapaume preserve some of the original landscape, albeit recontoured, sanitized or simply left for nature to reclaim.

Only a handful of British historians have successfully risen to the challenge of offering a credible, convincing and genuinely empathetic reconstruction of the 'everyman' experience, in or out of the trenches, on land or in the air.[2] In recent years the critical and commercial success of Sebastian Faulks and Pat Barker suggests that the novel is still the most potent and powerful means of depicting life in the trenches.[3] Outside of fiction, is it possible to recreate, for example, the noise, tension, aggression, and fear that was St Omer aerodrome in the spring of 1918, with the British Expeditionary Force (BEF) retreating and regrouping in the face of Germany's last great offensive of the war? To be fair, Nigel Steel and Peter Hart's *Tumult in the Clouds* offers an admirable overview of winged combat above the Western Front, plundering the oral archives of the Imperial War Museum to capture the physical sensation and emotional cost of waging war at 15 000 feet. The success of a book so firmly rooted in personal testimony confirms that there is simply no substitute for the memories and writings of those who were there.[4] In *Sagittarius Rising*, an acknowledged classic of aviation literature, Cecil Lewis recreated dawn at Clairmairais, or a dozen other airfields in northern France: the hangover, the layers of wool and sheepskin to ward off the cold, the smell of dope, castor oil and varnish, the yell of 'Contact', and the mechanic's swing of the propeller. The Wolseley Viper engine rumbles and roars as the chocks are pulled away – joystick gently back, throttle down and the SE5a scout starts to taxi down the field.[5] If, as Joni Mitchell once insisted, 'The drone of flying engines/Is a song so wild and blue' then it has to be left to a Cecil Lewis, or from another war, a Hillary or a Saint-Exupéry, to capture both lyric and melody.[6]

Mick Mannock was never given that same opportunity. Unlike his friend James McCudden, he died before the chance came to write about the war in the air. Across the decades since his death a succession of admirers and former comrades have sought to fill the gap, telling the rags-to-riches story of Edward 'Mick' Mannock, air ace and master tactician. First there were fellow pilots Ira Jones and William MacLanachan (writing as 'McScotch'), and then 30 years later a naval officer turned actor called Vernon Smyth joined forces with freelance writer Frederick Oughton. Finally, in the early 1980s a South African student of the First World War, James M. Dudgeon, fulfilled a longstanding ambition to chronicle his hero's achievements.[7] In between these four biographies numerous magazines and newspapers at one time or another ran features on the so-called 'ace with one eye'. Not that any of these writers were short of material to work from. RFC pilots were on the whole well educated.[8] Not surprisingly, to relieve long periods of boredom between operations they wrote a great deal – about themselves and about their fellow pilots. As well as compiling reasonably full combat reports after every mission, Mannock maintained regular correspondence with a wide range of friends, old and new. Between April and September 1917, while serving his apprenticeship with 40 Squadron, he also kept a diary, albeit not always on a daily basis.[9] Thus, both from his own papers where they still exist and from what other pilots were writing about him, it is possible to compile an almost day-by-day account of Mannock's 18 months in combat. He served on the Western Front from April 1917 until his death on 26 July 1918, during which time he rose from 2nd lieutenant to (acting) major and squadron commander, and was awarded the DSO and two bars, and the MC and bar. Released from Turkish internment he had rejoined his Territorial ambulance unit in July 1915, later sought active service as a sergeant and officer-cadet in the

Royal Engineers, and in August 1916 transferred to the Royal Flying Corps (RFC – from 1 April 1918 subsumed within a third armed force, the RAF). Officially credited with 50 victories, Mannock's unofficial score was posthumously calculated as 73, and on 18 July 1919 he was awarded the Victoria Cross.[10]

Mannock's wartime career would inevitably generate an abundance of material from which to retell his wartime exploits. However, the temptation, for both those biographers who knew him, and for their successors, was to look beyond the evidence, and actually reconstruct – or, to be more accurate, construct – his life. In other words, to tell their hero's story, whether or not it was verifiable. Thus, on almost every page 'McScotch' recreated conversations from 15 years earlier, while Smyth and Oughton conjured up pages of dialogue in a heroic if hilarious attempt to chronicle Mannock's early years. In their more honest endeavours even Jones and Dudgeon occasionally fell into the trap of writing 'faction' in order to fill in the gaps. The truth is that even the most assiduous researcher encounters problems in piecing together the life of an Edwardian working-class man for whom there is precious little on record. Of course once that man is no longer a civilian, and especially if he receives a commission, then wartime bureaucracy provides at least a partial rescue from the condescension of posterity.

When I first set out to write this book I was aware that any previous discussion of Mannock's politics had invariably been based upon hearsay, vague recollection, second-hand evidence and, above all, mere speculation. Nevertheless, I naively assumed that the papers of the Wellingborough Independent Labour Party (ILP) would be extant, thereby giving a unique insight into the day-to-day life of its prewar secretary, the then 'Pat' Mannock. Countless letters, telephone calls and e-mails revealed only that at some point in the past these records had been destroyed.[11] This revelation also stymied any notion of writing a 'definitive' life that would fill in all the gaps and correct all previous biographies' inaccuracies. To some extent I could still do the latter, but at the same time I had to be realistic and accept that there is only modest evidence upon which to build up a clear picture of how by 1914 Mannock had become a committed socialist, a convinced Home Ruler and a key figure in his adopted town's Labour movement. Chapter 2 nevertheless endeavours to provide just such an explanation, but resists the temptation to create wholly fictional scenes when recounting Mannock's childhood in an Indian garrison, his adolescence in Canterbury and his life in Wellingborough while a rigger with the National Telephone Company for whom he had previously worked as a clerk in Kent.

Similarly, Chapter 3 does not provide a graphic portrayal of Mannock the expat linesman languishing in an internment camp in Constantinople, eager to enter the fray once he could secure passage back home. Just as in the rest of the book, when writing about his stay in Turkey the emphasis is upon what we do know as opposed to what we don't. This piecing together of the course of events, drawing upon scraps of precious information gathered in the unlikeliest of places, contrasts starkly with our intimate knowledge of Mannock's experience once in France. As such it raises questions about the responsibility of the biographer when faced with an absence of evidence as to what was taking place at any particular point in time.

This book therefore is as much about the legend as it is about the man himself. It assumes that the life and the myth each warrant attention, if only because both raise so many questions about memory, biography, patriotism, commemoration, heroism and the 'other lives' of supposedly great men and women away from their public.[12] Why does Mannock still attract so much interest and admiration, and why should commemoration of his life and achievements be laced with controversy, not least concerning how he died and where he is buried? Why did the two RFC veterans who flew with him, Jones and MacLanachan, decide in the early 1930s that the time was right to mythologize Mannock? How did such a deeply unconventional person, the atypical English fighter pilot in terms of attitude, age and upbringing, come to be remoulded as a symbol of courage and patriotic endeavour, and a fine example to any future generation threatened from across the Channel – or to be more precise, the North Sea? The Mannock legend clearly owes a debt to the interwar fascination with the aviator, and his ideas impacted upon the next generation of fighter pilots, as confirmed by a close reading of Richard Hillary's *The Last Enemy*.[13]

Like Albert Ball, Mannock clearly appealed to a postwar public desperate for heroes free from association with all the worst facets of trench warfare. But this raises the question of how, had he lived, would he have dealt with fame and opportunity? The Second World War produced a new set of air aces, the likes of Bader, Gibson and Cunningham eclipsing the fame of Ball and Bishop, Mannock and McCudden. Given that the 1950s was a decade in which British cinema and popular biography was obsessed with the nation's 'finest hour', what prompted Vernon Smyth, advised by Frederick Oughton, to write his long and unintentionally amusing life of a by then half-forgotten hero? Jumping a generation, why do lifelong admirers like James Dudgeon still feel the need to go into print – is it to redress some vague grievance concerning their man's apparent absence from the pantheon of national heroes? As the final chapter reveals, Smyth's original intention was to make a feature film about Mannock. This prompts the question of how embedded the Mannock legend is in fictional treatment of the war in the air, whether in novels and films, or even on the radio.

The concluding chapter also asks, what are the purposes of commemoration, and why does a community – in this case, the two communities of Canterbury and Wellingborough – perpetuate the memory of a local hero? In recent years attention has begun to focus upon the dual process of remembrance and commemoration: the formalization of bereavement.[14] As the distance between 1914–18 and the present day grows ever greater, so this dual process increasingly has to take on board the remoteness of what we are being asked to remember, and the degree to which, in Jay Winter's words, 'objects invested with meaning related to loss of life in wartime become something else'.[15] Memorials retain a role in maintaining national, civic, and community pride, and yet for all but a very few their healing function, with regard to the First if not yet the Second World War, is to all intents and purposes complete. Similarly, the non-ceremonial memorials, such as a church hall or a cottage hospital, no longer fulfil their original purpose of enabling those who survived to remember those who did not by dedicating themselves to 'doing good'. Are memorials, 'artefacts of a vanished age, remnants of the unlucky generation that had endured the carnage of the Great War', or can they continue to operate as 'collective symbols', albeit over time assuming different functions from those for which they were originally intended?[16] The example of Mick Mannock would suggest this to be the case, with commemorations of his life fulfilling a variety of purposes, almost all of them complementary.

The two central chapters, covering Mannock's service on the Western Front, explore what might at first seem bizarre, namely that the secretary of the Wellingborough ILP's collectivist thinking directly influenced his approach to waging war in the air. In other words, surreal though the notion may be, he applied a socialist perspective to the destruction of enemy aircraft, leaving a legacy that survived until the Battle of Britain.[17] Mannock was a major influence on how aerial combat evolved in the final 18 months of the war as experienced pilots rapidly adjusted to new tactics and new technology. Inexperienced pilots were dead, either killed in training or in those first crucial weeks in France when not even good luck and a capacity to learn very fast guaranteed survival (in 'Bloody April' 1917 the life expectation of a newly arrived pilot dropped to just 17 days, and a year later front-line squadrons ground-strafing the

German advance were suffering an average of 30 per cent casualties *each day*).[18] Chapter 4 explains how Mannock survived his baptism of fire and overcame a debilitating fear of death: unpopular and open to a charge of cowardice, he quite deliberately avoided combat when the odds were heavily stacked against him, making good use of an extended apprenticeship to evolve a fresh approach to formation fighting, and to ensure both guns and aircraft performed to their maximum. Like fellow ace James McCudden, whose service career began in the ranks as a mechanic, Mannock understood the technology. By not relying on ground crew, and therefore taking a direct interest in the machinery and the weaponry, he could ensure a Nieuport 17 flew to its optimum. This willingness to get their hands dirty in the interest of long-term survival made Mannock and McCudden even deadlier in the skies over Flanders and Picardy once they were given an opportunity to boost the standard performance of the SE5a – they could fly faster, further and higher, and match similarly experienced pilots flying the last great German fighter of the war, the Focker D VII biplane.

The SE5a, perhaps more than any other aircraft of the period, embodied the scale of Britain's commitment to the war in the air by 1917, and the level of technological development achieved in the three years since an embryonic RFC (four squadrons) flew its motley assortment of fragile reconnaissance aircraft to France in the late summer of 1914. Of the 99 squadrons operational on the Western Front at the end of the war 14 flew the SE5a, of which (including the original, much inferior SE5) no fewer than 5205 aircraft were built. With over 30 000 aircraft delivered to the RAF in the final year of the war, and machines as sophisticated as the Vickers Vimy (capable of crossing the Atlantic less than a decade after Bleriot struggled to reach Kent) entering front-line service, the advances in production methods and engineering capability since 1914 were simply staggering.[19] Admittedly, from the perspective of the late 1930s, let alone the present day, a 1918 fabric-skinned biplane appeared crudely constructed, with a heavy dependence upon pre-industrial materials and a system of flight control both elementary and elemental. Nevertheless, comparison of the SE5a's 'office' with a prewar cockpit readily confirms that the means by which its pilot could gauge and control his machine's performance had radically improved. Most crucial of all for an interceptor aircraft like the SE5a, the combination of a greater understanding of aerodynamics (often acquired through bitter experience) and far more powerful engines such as the 200 horse power Viper had resulted in aeroplanes capable of flying at high altitude for long periods of time. Greater reliability, better manoeuvrability, increased speed and enhanced operator control combined in the second half of the war with a major leap forward in weapons technology that transformed the nature of aerial combat: the Allies' Constantinesco synchronizing gear prevented fuselage-mounted machine guns from firing when a propeller blade passed in front of the barrels.[20] The SE5a lacked the heartstopping manoeuvrability of the Sopwith Camel, nor did it capture the imagination of the general public in the same way – the Camel is to the First World War what the Spitfire is to the Second. But an SE5a flew faster and further, and with engine tuning courtesy of a Mannock or McCudden had the capacity to cruise at over 20 000 feet. With a fuselage-mounted Vickers .303, and a Lewis machine gun firing over the top wing, the SE5a has been coldly described as a 'steady gun platform', sturdy and deadly.[21] If any aeroplane embodied a four-year revolution in technology, and thus the means of destroying the maximum number of enemy aircraft with the least cost in material and manpower, then it was this ugly but potent product of the Royal Aircraft Factory and its partner companies scattered across Surrey and the Midlands. Mass-produced by a combination of unskilled – often female – airframe fitters and skilled toolmen, and then flown by 'temporary gentlemen' like McCudden, the SE5a symbolized the transformation that had taken place in Britain once the demands of twentieth-century warfare began to be fully appreciated.

The experience of Mannock and McCudden suggests that working- and lower middle-class pilots with a practical grounding in basic engineering had a better chance of surviving for an extended period of time in France than young officers recruited straight from school or university with a classical/liberal arts education. As will become clear, Mannock and McCudden had very different personalities, with the former considerably older. Yet both were sons of soldiers and 'Men of Kent'; and both knew what it was like growing up in a working-class Irish Catholic family with several siblings and precious little money coming into the home. Their attraction is precisely because they are so very different from the stereotypical fighter pilot, by virtue both of their humble origins, and, in consequence, their ruthless approach to waging war. Their upbringing, education and early experience with the harsh and dirty Edwardian world of work left them indifferent to any lingering notions of chivalry, and cynical towards any suggestion that morale rather than machinery would ensure aerial supremacy deep into German-held territory. How refreshing therefore that in 1997 the Imperial War Museum's Nigel Steel and Peter Hart dedicated *Tumult In The Clouds: The British Experience of the War in the Air, 1914–1918* to Mannock and

McCudden, alongside a reminder from John Lennon that, 'A Working-class Hero is something to be'.[22]

What happens to a 'Working-class Hero' after he returns home from doing his bit? In the case of Mannock and McCudden we shall never know – both died in the summer of 1918. Mannock never speculated upon what the return of peace might bring as he felt certain he would not be alive to enjoy it. Yet his failure to survive the war has never stopped his admirers from speculating upon what he might have achieved in peacetime. The normal assumption is that he would have risen to the top in the Labour Party. But would he? What other courses of action were open to him in the aftermath of the 'war to end all wars'? Might he, for example, have followed the example of another one-time RFC officer, and transferred his allegiance to Sir Oswald Mosley's Black-shirts? Alternatively, might he have joined other left-leaning fliers eager to fight fascism in defence of the Spanish Republic? We know what happened after the war – Labour politics was denied the glamour of a genuine war hero like Mannock – but need this prevent us from exploring events 'in the context of what might have happened'?[23]

Mannock died four months short of the Armistice, but what was the experience of Labour activists who had 'a good war' once they re-entered the peacetime political arena? In John Banville's masterly study of deception and deceit, *The Untouchable*, his 'fifth man', Victor Maskell, reflects on the vacuity of the phrase 'a good war', suggesting that it is no more than a dramatic conceit:

> In the films of the late forties and the fifties, pomaded, soft-faced chaps in cravats were always pausing by the fireplace to knock out implausible pipes and ask over their shoulders, 'Had a good war, did you?' at which the other chap, with moustache and cut-glass tumbler from which he never drank, would give one of those very English shrugs and make a little moue of distaste, in which we were supposed to see expressed the memory of hand-to-hand combat in the Ardennes, or a night landing on Crete, or a best friend's Spitfire going down in a spiral of smoke and flame over the Channel.[24]

This of course was exactly how Vernon Smyth saw Mick Mannock, and why in the late 1950s he wrote a screenplay which at the beginning and end has veteran pilots like Ira Jones revelling in understatement in the nave of Canterbury Cathedral – a case of right decade, wrong war. Not surprisingly, therefore, when Smyth co-authored a biography of Mannock he displayed a total misunderstanding of what made the man tick.

Held together by Scotch and an increasingly incoherent hatred of the Germans, and haunted by nightmares of going down in a 'flamer', the burnt-out ace who returned to France in June 1918 certainly wasn't 'good old Mick' – near the end of his life nearly all of the most attractive facets of his character appear to have been literally shot away. Thus, the notion of 'a good war' as discussed in the final chapter of this book is not the stiff upper lip and the rows of medal ribbons. Having 'a good war' is simply surviving front-line combat with one's dignity reasonably intact, and a quiet satisfaction at sharing in ultimate victory, albeit at a terrible – and some would say, unacceptable – cost. Applying such straightforward criteria, the majority of those who served on the Western Front, or any other theatre of operation – even those most tortured by post-traumatic stress – can be judged to have had 'a good war'. Seen in the context of the Labour Party, it's Clem Attlee recalling Suvla Bay, 'Henry Dubb' Tawney reliving the first day of the Somme, or Hugh Dalton revisiting Alpine artillery emplacements.

Discussion of what role Mannock might or might not have played in interwar Labour politics prompts a wider consideration of the party's uneasy relationship across the century with matters military. Was the experience of Labour politicians returning from war after 1918 different from those resuming or commencing political careers in 1945? What are the particular difficulties and tensions Labour leaders experience when in office at a time when British forces are in action overseas, particularly when they themselves have not been in uniform?[25] Ironically, the interwar premier most enthusiastic about the civil and military potential of aircraft and airships was Ramsay MacDonald, principled opponent of war in 1914 and oft-proclaimed 'Champion of the League of Nations'.[26]

MacDonald had an inexplicable, and wholly uncharacteristic, affection for piston-engined aircraft, which I share.[27] My father worked in the aviation industry, and like so many little boys in the 1950s and early 1960s I assembled Airfix kits and collected the *Eagle*'s cut-away diagrams, always preferring Hurricanes and Spitfires to Hunters and Scimitars. That early fascination with a pre-jet era almost inevitably focused upon the Second World War, but as the years passed and boyish enthusiasm gelled with academic interest I found myself increasingly intrigued by an earlier generation of aviators. Chapter 2 briefly recalls the first time I heard of Mick Mannock, in my final year at primary school. In my first term as an undergraduate in Canterbury I discovered his memorial tablet in the Cathedral, thus proving to a friend's wife born and raised in the city that her great-aunt was by no means senile in insisting that a long-dead brother-in-law had won the VC flying in the Great War. The next time I thought seriously about Mannock was in 1997, when I came across a reference to him as a socialist. Intrigued, and attracted by the thought of being able to combine my amateur interest in aeroplanes with my professional interest in the Labour Party, I decided to find out more about the only fully paid-up member of the ILP to command a front-line squadron on the Western Front. Only later did I discover that Mannock was a fellow Catholic, with a similarly relaxed view of doctrine and adherence to the one true faith.

Researching this book, arising out of an early, and thankfully unpublished, essay on Mannock's politics, was a labour of love, not least because it necessitated several visits to Kent, the discovery that Wellingborough is by no means a demoralized victim of deindustrialization, and the opportunity to combine archival work with R & R in the halls and galleries of the RAF Museum, the Imperial War Museum and the Museum of Army Flying. My thanks to the archivists of those august institutions and of Canterbury Cathedral, and of course library staff at the University of Southampton, and also to Matthew Midlane, who facilitated access to regimental histories in the library of the Royal Military Academy, Sandhurst. Footnotes throughout the text provide acknowledgement of specific assistance, but particular mention must be made of the support provided by friends and fellow historians at the former University of Southampton New College, particularly Commander Paddy Johnston who read and commented sagely upon everything I have written about Mannock, and Derek Edgell who very kindly took over my administrative duties in the autumn of 1999. My former research student, Alan Brown, took time off from his own work to provide early advice on sources and use of the Internet. Elsewhere in the university, Bob White demonstrated the technical expertise of a professor in aeronautics and the enthusiasm of a lifelong student of the war in the air, and Jon Clark ensured I had the time and opportunity to complete this project. Nearer home, Adam Smith designed the map of the Western Front, and Edward Read displayed similar computer wizardry in ensuring camera-ready copy. However, no serious study of 'Uncle Eddie' is possible without the assistance of his niece, the late Jeanne Ogilvie, and her daughter, Anne Tooke. I am especially grateful to them for their hospitality, their good company, their never-ending flow of material and suggestions, and above all, their patience when my ignorance became all too obvious. Finally, as always, a very special thanks to Mary and Adam for tolerating yet another project, and for showing such patience as we drove around the back streets of Wellingborough or the suburbs of St Omer in search of Mick Mannock.

2
A Prewar Education, 1888–1914

From Cork to Canterbury

Six miles west of Cork, Ballincollig today is a satellite town, with the barracks on one side of the road and the 'village' on the other, the latter largely comprising breezeblock estates and small, cheap, rather tacky – but well-intentioned – shopping malls. Ballincollig's past can be resurrected courtesy of two cardboard boxes tucked away in a corner of the community centre's crowded library, and a visit to the gunpowder mills' heritage centre. Gunpowder brought employment to nineteenth-century Ballincollig, and it also brought soldiers. In 1867 fears of Fenian sabotage saw the garrison reinforced, and the British Army remained in strength until its withdrawal in May 1922: intent on wiping out all vestiges of foreign rule before Free State forces could move in, Anti-Treaty irregulars then proceeded to burn down the barracks.[1] Life in Victorian Ballincollig revolved around the needs of the soldiers and the 500 or so workers and their families housed in the mills' tied cottages. By the end of the century the inability of black gunpowder to compete with dynamite had led to a significant decline in the workforce. However, train and tram links with Cork, and a cavalry regiment permanently garrisoned there, had seen the local population rise to well over 2000, enough to support eight pubs and a large Roman Catholic church dating from the 1860s.[2] It was in this imposing neo-Gothic church – today surrounded by council houses and a carpet factory, but a century ago high on a hill above the village – that on 4 February 1883 Julia Sullivan married Corporal Edward Corringham. A cavalryman from England on a two-year posting to Ballincollig with the Royal Scots Greys, Corringham had taken an assumed name.[3] In an era when peacetime soldiering was viewed by respectable society as an occupation fit only for scoundrels, shirkers and soaks, a comfortable Victorian household would be loath to find the family name resonating across the parade ground at Aldershot or Chelsea. A decade later, when Corringham left the regiment, he reverted to his real surname – Mannock.[4]

Corringham/Mannock was the youngest son in a large family. He came from London, where his father appears to have worked in Fleet Street; but the family had strong Irish connections, his parents having met and married in Dublin. If truth were told precious little is known about the Mannocks, for all the speculation of earlier biographies. They are only one of many semi-anonymous rising middle-class families lost in Victorian suburbia, a challenge to even the most diligent genealogist. The exception is J. P. Mannock, Corporal Corringham's next eldest brother, and a shadowy, raffish character who is said to have taught billiards to the Prince of Wales. Whether or not John P. graced the green velvet of Sandringham and St James's, he was happy to reveal the secret of his success in *Billiards Expounded*. This guide to the game appeared in 1904, by which time Mannock's playing career was well and truly over. Throughout the 1890s, in salubrious gentlemen's saloons around The Strand and Soho, the waistcoated and white-tied maestro of the imperial game surrendered his crown in a succession of high-profile matches, culminating in humiliation at the hands of the German champion in March 1897 and again in October 1898.[5]

Heavy betting and hard drinking at the Argyle Street Billiard Hall or the Dean Street Saloon was a long, long way from soldiering – and courting – in west Cork. The Sullivans were a well-established family in Ballincollig, providing some of the first labourers employed at the gunpowder mills. Julia herself worked in the post office, and her parents were servants on a large local estate.[6] Coming from such a poor background, marriage to an English corporal with apparently rich relatives seemed at first sight to offer an escape from poverty: married quarters, however cramped and basic, had to be an improvement on a damp tiny cottage sparsely furnished. Although army pay was low by English standards, with a corporal earning no more than two shillings a day, wages were nevertheless higher than could be earned on an Irish tenant farm or estate. In addition, deferred pay of two pence a day ensured a lump sum when the soldier eventually returned to civilian life, and in Corringham's case this meant he could anticipate receiving around £40 upon discharge from service.[7] Yet clearly her parents disapproved of the match as no relatives attended the wedding, even though Corringham had accepted that marriage was out of the question unless he became a Catholic: he had been received into the Church on 16 January 1883.[8]

To convert to Roman Catholicism in the late nineteenth century was not a step to be taken lightly given the level of suspicion and ignorance still evident in even the politest of circles. The Royal Scots Greys were overwhelmingly Protestant, and once the regiment left Ireland then Corringham and his new wife would encounter a succession of practical problems (where to hear Mass? where to educate the children?) let alone a gut prejudice towards 'Papists'. The chances are that, like Henri IV, embracing Rome meant no more to Corringham than a means to an end. However, for his wife and children the family faith proved both a constraint and a comfort, a burden and a badge.

Although some reforms had been initiated, not least in improving crowded and unhealthy accommodation, for privates and NCOs alike barracks life was cruel, crude and harsh. Preparation for an eventual return to civilian life was non-existent, with scant prospect of a decent job awaiting even the most diligent and temperate soldier. The latter was an especially rare phenomenon, with strong drink a universal relief from the tedium of peacetime soldiering. Marriage was discouraged, with wives forced to embrace a bruising lifestyle in an unfamiliar environment cut off from the wider world. Isolation from civilian life was compounded by husbands' inability and indeed reluctance to accommodate domesticity, the mess invariably more attractive than the married quarter.[9] The regiment's wives shared a common identity and a common, all too often miserable, experience; but loneliness and insecurity was compounded by the upheaval of fresh postings: the 2nd Dragoons of the Royal Scots Greys moved five times between 1883 and Mannock/Corringham completing his service nine years later. For much of that time the regiment moved around Ireland, if only because of easy access to the Curragh for cavalry exercises; in addition, the Royal Irish Constabulary appreciated mounted troops being readily available in the event of civil disorder, such as the Belfast riots in 1886 over the first Home Rule Bill.[10] Summer exercises meant husbands could be away from their wives for several months. During one such absence Julia gave birth to Jess, the first of two daughters. Patrick arrived in June 1886, by which time the now Sergeant Corringham and his family were based in Aldershot. Two years later, on 21 May 1888, at Preston Cavalry Barracks in Brighton, a second son was born, Edward.[11]

Mannock's most recent biographer, James Dudgeon, insisted that family papers in his possession showed Cork to be the real birthplace. There is of course no reason why Julia, with two toddlers to look after, might not have wanted to return home for her third confinement. She was by all accounts a formidable woman, and her relationship with

Corringham was stormy, to say the least. However, Dudgeon placed Julia in Cork because the 2nd Dragoons had been posted there, and the records confirm that in both 1888 and 1887 the regiment remained in the south of England.[12] If 'Pat' Mannock was an Irishman, which, as Home Rule returned to the top of the Edwardian political agenda, he increasingly declared himself to be, then it was almost certainly through kith and sentiment, and not an atavistic attachment to the land of his birth.

His term of service completed, Corringham returned to civilian life in 1891 under his old name, the Mannock family taking rented accommodation in Highgate, north London. Most of the £40 lump sum went on whisky and porter, and employment prospects were bleak given the generally low esteem in which ex-soldiers were held. Mannock found living in cramped rooms with a resentful wife and three young children a feeble substitute for life in the cavalry. A failure to adjust to fresh circumstances, and the loss of what little authority he had previously enjoyed on the parade ground, left him rootless, moody and resentful. His family suffered the consequences of a foul temper and a ferocious thirst. After 18 months of misery Mannock walked in to a Liverpool depot and rejoined the Army, as a trooper in the 5th Dragoon Guards.

Posted to India, and swiftly promoted on the basis of past experience, Corporal Mannock arranged for Julia and the children to join the regiment at Meerut, north of Delhi. Notwithstanding the intense summer heat, married quarters for an NCO and his family in India were usually more comfortable than at home, if only because cheap labour and a favourable exchange rate facilitated employment of at least one servant. Married cavalrymen saw far more of their families given that for much of the year they were required to do little more than parade and groom their horses, and play sport on Thursdays and Sundays. Training was generally restricted to the cold weather months between October and March, so that prolonged absences from home were rare. Lengthy separation was usually the result of wives and children seeking relief from the summer heat in the northern hill stations. Sleep and drink no doubt took priority, but ironically Mannock found himself spending more time with his children than he had as a civilian. This leisurely lifestyle prevailed for five years, until Robert Baden Powell assumed command of the 5th Dragoon Guards in 1898 and gave NCOs a key role in pioneering fresh training methods and tactics.[13] A year later, in September 1899, Mannock sailed with his regiment for South Africa and the second Boer War. He was besieged at Ladysmith, chased De Wet's commando across the veldt, and later aimlessly criss-crossed the Transvaal in a succession of mobile columns and baggage trains.[14]

Astonishingly, when the regiment was ordered back to India in March 1902, Mannock opted to serve out his time in South Africa rather than be posted back to Britain. He transferred to the 'Black Horse', the 7th Royal Dragoon Guards, in May 1902 taking part in the last major battle of the war. Mannock at last returned to England, and a final posting at the Cavalry Depot in Canterbury.[15] It was here – at 19 Military Road, a small terraced house adjacent to the barracks – that slowly and very painfully he became reacquainted with the family he had not seen for three years. He had shown precious little interest in Julia and the children while away fighting the Boers, and within only a few months of their being reunited he walked out for the last time. Free from army discipline, Mannock chose to cut loose, seeking fresh opportunities and fresh women.[16]

Young Edward spent nearly a decade in India, and the truth is that little is known of his time in Meerut other than what has passed into family folklore. He was it appears a quiet, sensitive child, keen on books and birds, and increasingly distant from a scornful father in a permanent rage. Not unnaturally, as the years passed, and particularly after Mannock sailed for South Africa, Edward drew ever closer to his mother, and was avowedly her favourite. Like other garrison children not destined for private schooling at home, Edward received an elementary education up to the age of 14. However, his schooling was courtesy not of the Army but the Society of Jesus. The Jesuits had been well established in India as a missionary and teaching order since the late sixteenth century, but – unlike Kim – Edward never graduated to the prestigious St Xavier's of Lucknow. Nevertheless, his catechismal grilling was presumably more rigorous and testing than any available from a Father Victor (only Irish infantry regiments like Kipling's fictitious Mulligan Guards enjoyed reassurance from Catholic chaplains that the Holy Trinity *and* the Virgin Mary were unquestionably on their side).[17] As will become apparent, Mannock's relaxed brand of Catholicism never wholly disproved Loyola's claims for the Society's success rate in moulding young minds.

Ironically, the ignorance surrounding Mannock's childhood in India prompted earlier biographers to write page after page of speculation and pure fiction. Frederick Oughton and Vernon Smyth managed to fill 20 pages with a succession of scenes lifted from their abortive film project: biographical narrative was reduced to speculative dialogue, and very entertaining it is too – albeit for all the wrong reasons. Even the conscientious James Dudgeon succumbed to the temptation to fantasize on a dysfunctional Mannock family permanently on edge in the heat of a north Indian summer.[18] This soap opera style of history focuses upon three principal themes: Mannock's alienation from his father, his early sense of injustice and his impaired vision. Implicit in this fascination with Mannock's early childhood experience is the assumption that these are formative years – an assumption rooted only in hearsay and dim recollection.

My earliest acquaintance with 'Mick' Mannock was courtesy of a 1962 BBC Schools drama production on the life of the 'ace with one eye'.[19] A one-eyed air ace is a wonderful idea worthy of W. E. Johns, albeit not quite as potent and resilient an image as flying with no legs or a reconstructed face. Neither Douglas Bader nor Richard Hillary could hide their scars, and both wrestled with authority before securing permission to fly again. Ultimately, their propaganda value was realized, with Bader, even while still languishing in a POW camp, fast becoming a living legend. Hillary's is a more complex story, to which this essay must return. Mannock was supposedly able to keep his impaired vision a secret, confiding in only a few close friends for fear of being grounded. Flying and fighting with only one good eye generated paranoia not pride, its propaganda potential being the last thing on his mind. This assumes, of course, that Mannock had impaired vision, and that he did indeed suffer from some form of optical astigmatism in his left eye.

There is a strong temptation to dismiss the notion of the 'ace with one eye' as convenient myth-making, guaranteed to enthral little boys brought up on a 1950s diet of *Reach For The Sky* and 'Bandits at 12 o'clock high!'. How could a front-line pilot survive 17 months' near-continuous combat with a complete loss of peripheral vision on one side?[20] Mannock's service file records medical examinations on 22 May and 24 November 1915, and 2 August 1916. On each occasion he was found to have perfect 'acuteness of vision' in both eyes.[21] Yet, even if we dismiss earlier biographers' flights of fantasy as to what occurred in India, their sources within Mannock's immediate family were insistent that as a child he had contracted a seriously debilitating amoebic infection, one consequence of which was to render him far more introspective and sensitive. In other words, if not actually traumatized by the experience of temporary blindness, young Edward emerged from his experience less thick-skinned and also less indifferent to the harsh vagaries of life than most of his fellow service offspring. Of course this does not necessarily mean that his eyes were permanently damaged, and indeed Dudgeon quotes one close relative's insistence that Mannock's 'sight was no worse than most people's; in fact it was much better than average'.[22]

Yet back in the 1930s both Jones and MacLanachan had insisted that their hero flew with a congenital defect in his left eye. MacLanachan rates only a passing reference in Ira Jones's book so it is unlikely the latter was familiar with the anecdotes recounted two years later in *Fighter Pilot*. In other words, Jones had heard independently the story that Mannock fooled the medical officer by using a delay in testing his eyes to memorize all the letters on the Snellen chart. This was the story which, told by the BBC almost half a century later, so captivated me that I never forgot it. MacLanachan's slightly different version has Mannock tell him that the doctor, 'covered my bad eye first and gave me time to make sure of the letters before he covered the good one'.[23] This sounds a pretty tall story, except that apparently the right eye is usually tested first, so that anyone with a photographic memory could in theory pass the test. Also, memory tests were a favourite Victorian/Edwardian parlour game, with the Mannock family no exception.[24] But, even if such instant recall was feasible could Mannock have carried it off on three separate occasions?

MacLanachan claimed that Mannock let him into the secret after his Nieuport hit a haystack for no apparent reason: hitting the cockpit with his 'good' right eye while crash landing had rendered him effectively blind. This crash certainly occurred, as did an earlier incident when Mannock lost vision in his right eye. However, it is hard to reconcile MacLanachan's claim that Mannock confided to him, 'I can't see on that side; but it's all right, I can still see with my other one. I thought *that* was done in,' with the fact that on the previous occasion he had flown the following day. With his right eye heavily bandaged and feeling 'like a bell tent', how would he have seen to fly if he had been nearly blind in his left eye? Dudgeon makes the same point, and with refreshing scepticism casts doubt upon 'McScotch's' story, quoting letters to him from Keith Caldwell and H.G. Clements – Mannock's CO and wingman in 74 Squadron – that no one who was such a good shot could have had poor eyesight.[25] The RAF in the summer of 1918 was still sufficiently small and intimate that stories and rumours concerning Mannock would have been widespread in the weeks following his death. The myth-making had begun even while he was still alive, and what better way of enhancing one's claim to have been on intimate terms with the great man than to claim that the real reason for his unusual stare was a partial blindness? Jones's book makes no mention of Mannock's impaired vision in connection with combat, so presumably when they flew together it was never an issue. However, this does not mean to say that he would dismiss the story if he heard it after Mannock's death – and clearly he did not. By 1934 Jones was insisting that impaired vision was 'noticeably traceable in a number of men grown to fame and of outstanding genius in history', but regrettably refrained from quoting examples. The temptation to rely on rumour rather than fact must have been too great, and with 'McScotch' then quoting the man himself, the legend of the one-eyed ace quickly became 'reality'. A few sketchy details about what took place in India around the turn of the century became inflated into the tale of a plucky but deprived lad who survived the terrible test of blindness to – quite literally – see the world from a fresh perspective.

Similarly, the unhappy relationship between Mannock and his father became inflated into a brutal domestic struggle between an innocent invalid, at one with nature and ever thirsty for knowledge, and an uncomprehending philistine whose preferred means of communication was his fist. A story of courage and heroism has to begin somewhere, so what better scenario than young injured Eddie's stoic endurance of paternal bullying? Mannock's indifference to parades and toy soldiers – so infuriating to his father – is of course ironic given his eventual aptitude for waging war. In the light of later events it can also be read as an early rejection of authority. Dudgeon prefers, however, to focus upon the poverty and privilege within the barracks – and beyond. This 'large sprawling country' captures young Mannock's interest, 'as a sponge draws water'. He sees that social divisions in the Indian community mirror those of the governing elite: inequality is either side of the *maiden*, the only difference is of degree. Thus, imperialism merely compounds a systemic, inherent injustice. All this is described in a florid fashion, albeit without the cod-Irish dialogue at the heart of Oughton and Smyth's tedious family drama. While previous biographers had no more than hinted at an early social conscience, Dudgeon saw the need to extrapolate from later events a young man intent on righting wrong.[26] The child may be father to the man, but if truth be told there is no solid evidence that the years spent in India were genuinely formative. On the other hand, Mannock was 14 by the time he left Meerut, old enough to leave elementary school, *and* to have gained some insight into the distribution of wealth and power between the 'haves' and the 'have nots', whether on the plains of India or – where he now found himself – in the garden of England. Arrival home must have been a major culture shock, compounded by the realization of just how poor the family really was, certainly by comparison with his well-to-do cousins in west London.

For soldiers and civilians alike, Edwardian Canterbury was 'all parsons and pubs' – on the one hand England's premier cathedral city, and on the other a modest-sized market town in east Kent surrounded by hopfields and orchards. Apart from some rather grand late Victorian buildings on and off the High Street, and new villas for the aspiring middle classes south of the city, what was remarkable about Canterbury was how little it had changed since Dickens's day. Indeed, even allowing for the extensive damage inflicted by the *Luftwaffe* in June 1942, the streets surrounding the cathedral precincts only really began to take on a different character when redevelopment began in the late 1960s. The advent of mass tourism in the 1980s, and a burgeoning of the student population, saw the final disappearance of Canterbury as an identifiable community: for all the continuity of the cathedral's still imposing presence, ring roads, supermarkets and shopping malls all but destroyed the inner-city network of cramped streets, small shops and compact housing which for nearly ten years Mannock thought of as home.[27]

Once Julia Mannock was left to bring the children up on her own she moved out of the terraced house in Military Road into one of the neighbouring 'Jones Cottages'. Every Sunday the family would have walked through Northgate, and around the city wall, in order to attend Mass at St Thomas of Canterbury, the Roman Catholic church consecrated by Cardinal Manning in 1875.[28] Located in the Burgate, literally in the shadow of the cathedral, the church was a glorious act of defiance – a temporary arrangement until Rome's recalcitrant children saw the error of their ways.[29] At a time when even the most ecumenical Anglican viewed ultramontane Catholicism with deep suspicion, worshipping at St Thomas's must have been a lonely and dispiriting experience, particularly if hand-me-down clothes and an Irish brogue marked you out from the wealthy English patrons of this tiny, beleaguered congregation. Contrary to popular assumption, only Mannock's youngest sister, Nora, attended St Thomas's School, the remainder of the children being forced to find work as soon as the law allowed.[30]

Mannock's first wage was 2s 6d (12.5p) a week, earned by shifting and delivering greengroceries for ten hours a day. After a few months of retail servitude the exhausted teenager doubled his pay and eased his aching frame by taking a job in a barber's shop. He swept the floor, lathered shaving soap, washed hair and catered for each customer's every need. Lather boys worked unusually long hours, even by the standards of the 1900s. A barber's busiest times were evenings and weekends. The shop Mannock laboured in might have been open for anything up to 15 hours on a Saturday, in order to satisfy all those who only needed a shave once a week. It's not surprising, therefore, that he resolved to leave the beards, baccie and beer, let alone the boredom, at the earliest opportunity. Although well-qualified to join the 'black-coated workers' – he was bright, literate and, unusually for his age and background, well-travelled –

Mannock had been reluctant to accept a clerical post, hating the idea of being stuck behind a desk or counter. However, although he eventually grew to just over six foot and his body filled out accordingly, poor diet and recurring malaria left young Eddie a late developer. If he couldn't manhandle sacks of potatoes with consummate ease then farmwork appeared out of the question, especially if the hours were as long as in the barber's shop. In the end a combination of fatigue and family pressure resulted in him joining his elder brother Patrick as a ledger clerk for the National Telephone Company.[31]

Mannock had reached an age when 'messenger' jobs would begin to dry up. Shopkeepers required cheap labour, so anyone old enough to demand a man's wage could expect the sack. However grudgingly he accepted the offer of a lifetime counting columns, he must have recognized that, without the family connection, his chances of securing regular employment were slim. Although confined to a cramped office, and under close supervision, by Edwardian standards the conditions were by no means intolerable. Certainly, the pay was a distinct improvement on a lather boy's five bob (25p) a week plus tips. Furthermore, if he survived 35 years of tedium and toil there was a modest pension to keep him from penury in his dotage. Mannock stuck to accounting for three years, finally calling it a day in early 1911. He never came to appreciate the magic of double-entry bookkeeping, or relished the prospect of a job for life. Neither in fact did his elder brother, who only remained a loyal company servant out of a sense of duty and responsibility to the family. Edward still yearned to work outside, especially as winter damp and summer stuffiness rendered him ever more pasty-faced and bronchial. However, as time passed his ambitions extended beyond a simple yearning to work in the sunshine and the fresh air. Like Leonard Bast in *Howard's End*, Edward Mannock sat at his counting desk thinking of higher things. Not that the young clerk's hopes and aspirations were as elevated as Mr Bast, but then Mannock never met a Margaret Schlegel. He did, however, meet Cuthbert Gardner.

Cuthbert Gardner was only 26 when he first came across Mannock, recently arrived in Canterbury. The same year – 1903 – Gardner established his law firm in Castle Street, having been articled to solicitors locally, and for a brief period in London. With shares in the family brewery, Gardner was a moderately wealthy man even before his practice began to flourish. Although not publicly involved in party politics he represented the Westgate ward on the Board of Guardians from 1910, and after 1930 its successor, the Board of Assessment.[32] By the time of the First World War Gardner was already a pillar of the community, serving as first deputy and then city coroner for a remarkable 53 years. For much of that time he was also Clerk to the City Charities, a natural extension of his voluntary work for a variety of local organizations.[33] A bachelor, Gardner had two preoccupations: the law, and God. A warden at St Gregory's, the Anglican church only a short walk from the Mannocks' home on Military Road, Gardner ran the parish cricket club and the Church Lads' Brigade. He probably first got to know Mannock after the latter joined St Gregory's CC, progressing over the years from a bystander in the park begging for a bowl to captain of the Saturday team. Either as an act of evangelism or of philanthropy, Gardner encouraged Edward to don his first uniform and, as a drummer in the Church Lads' Brigade, to fight the good fight. The latter advice was to be taken literally as Gardner was not averse to livening up summer or Easter camps with wide games based upon military manoeuvres. Indeed, when a real war came along, the Church Lads' Brigade was absorbed into the city's Army Cadet Corps, with Gardner on the Reserve List for the duration. The ease with which he swapped uniforms in 1914, let alone the prewar enthusiasm for matters military, suggests Gardner was as happy bearing arms as bearing witness. If he could persuade Julia Mannock that a son sharing a tent with Anglicans did not mean eternal damnation, then he could easily convince Edward that Catholicism and patriotism complemented not contradicted each other. In other words, that the Church Lads' Brigade was an ideal apprenticeship for the newly formed Territorial Force.[34]

Gardner's enthusiasm for the Territorials was matched by that of Mannock's closest friend at the time, Fred Rawson. Fred's father was Albert Rawson, a staff sergeant in the 2nd Home Counties Field Ambulance, Royal Army Medical Corps (RAMC Territorial); but previously a 'rough rider' alongside Corporal Mannock out on the veldt with the 7th Dragoons. The Rawsons' house in King Street was only a few hundred yards away from Gardner's unusually modest abode in Palace Street. Fred's father not only suggested Mannock become a part-time soldier, but he invited him to lodge with the Rawson family. Fred and Eddie were 'like brothers', the latter abandoning his drum for the bugle when he joined the Field Ambulance unit's Transport Section.[35] Mannock must have inherited his father's horsemanship as he learnt fast: by February 1910 he was a full corporal running Friday night courses on 'care of harness and animal management'. On Monday nights at the Castle Street drill hall he paraded and practised first aid with around 170 other soldiers, NCOs and officers. Employers were obliged to release Territorial Force members, and every Thursday and Saturday afternoon Mannock found himself at the cavalry barracks absorbing the finer points of riding drill. He rode well, and although technically a non-combatant, photographs show him with a standard cavalry sword prominent in its saddle scabbard. Every July Mannock supervised the horses' entrainment to summer camp, and by 1912 his enthusiasm and aptitude for soldiering had earned him promotion to sergeant.[36] His commanding office was Major, later Lieutenant Colonel, A.R. Henchley. The Territorial Force had been a Liberal initiative, introduced as part of R.B. Haldane's army reforms of 1907, and Henchley was a keen supporter of the government. Not surprisingly, he was opposed to conscription, unlike many of the Unionist businessmen who dominated the corporation and would have been sympathetic to the National Service League's call for compulsory training.[37] Nor was Henchley's determination to turn his townies into real soldiers likely to impress local landowners who, despite the harsh lessons of the Boer War, still saw foxhunting as a mainstay of the East Kent Yeomanry's training for the next imperial war.[38]

Cuthbert Gardner's office was only two streets away from the National Telephone Company, and even when not meeting Mannock on parade or at the wicket, he could scarcely have avoided bumping into him on the High Street. Whatever his motives, and we must assume that they were wholly honourable, Gardner shared the enthusiasm of the Edwardian male ameliorist for helping working-class lads who wanted to get on in the world. The relationship evolved over the years, and by the war was presumably on a more equal footing, not least because once Mannock was commissioned he named Gardner as his legal representative. In old age Gardner boasted of his encyclopaedic knowledge of Mannock's life, pointing out to the local press that Oughton and Smyth's account might have been more accurate had they consulted him more.[39] Gardner's politics were typical of his class and background, and if he supported the Liberal Party then he was wise to exercise discretion in such a staunchly Unionist stronghold. Either directly or indirectly, he encouraged Mannock's engagement with a world beyond the Trollopian machinations and Tammany Hall politics of a city which, for all its diocesan pretensions, remained a provincial backwater.

Politics in Canterbury – a genuinely rotten borough, and with a tiny electorate in proportion to the total male adult population – largely focused upon control of the local Unionist Party. To all intents and purposes it was a pygmy power struggle, between long-standing Member John Henniker Heaton, and the Lord Mayor, Francis Bennett-Goldney. The latter stood as an 'Independent Unionist' against his party's official candidate in both elections in 1910, turning a first narrow defeat into a second easy victory. Although Hennicker Heaton was an enthusiastic tariff reformer, his critics resented too many speeches in the House on the case for an Imperial Penny Post, and too few on the need to protect local farmers.[40] In 1908 the Kent Tariff Reform League organized a series of demonstrations in Canterbury demanding a duty of 40 shillings on all imported hops. On several occasions large numbers of labourers and hop-pickers, most of them disenfranchised, demonstrated vociferously outside the Corn Exchange. Mannock would have stared down from his office window upon St George's Street, watching the crowd swell around a large cart packed with bowler-hatted speakers and furiously scribbling reporters.[41]

Tariff reformers in and around Canterbury regarded the city's fledgling ILP branch as an easy target: in Parliament, Labour MPs were firmly committed to free trade, not least because of Philip Snowden's unique – and ultimately fatal – combination of ethical socialism and Gladstonian Liberalism. In a constituency where even the Liberals struggled to survive, the Independent Labour Party lasted less than 18 months. Inspiration came from a young man in the Town Clerk's office, Walter Speed, who with missionary fervour sowed the seeds of socialism in the less than fertile soil of east Kent. The Canterbury branch quickly collapsed after Speed left in spring 1909 to become Labour's first full-time official in the Medway towns. The secretary, ever the optimist, had established his credibility within the party by organizing a succession of open-air meetings. ILP members in Dover, Ashford and Maidstone swelled the numbers and joined in the speech making. Nor were meetings restricted to municipal parks and half-empty halls. Speed was not afraid to venture into the surrounding countryside, rousing somnolent Dissenters in the villages between Canterbury and the railway town of Ashford. On the coast he founded a fresh branch in Deal, and pledged solidarity with a one-man 'Socialist Society' ('who would be pleased to hear from any speakers spending their holiday at Whitstable'). In September 1908 the Kent Federation of the ILP propagandized in Canterbury for a week, culminating in yet another rowdy meeting where speakers were shouted down by tariff reformers, many of whom were drunk. The big question of course is, was Mannock ever present when Walter Speed and his comrades struggled to make themselves heard above the hecklers? Did he ever hear the unflagging spokesman of the south's labouring classes make the case for 'why Christians should be Socialists', or proclaim the arrival of the 'Canterbury Labour Church'?[42] If, as his siblings later recalled, Mannock was by now an avid newspaper reader, he would surely have been aware of the ILP's activities in those heady days of 1907–9. Only two years later Speed was dead, already a fading memory in the terraced rows of Wincheap and Whitstable and the committee rooms of Dover and Deal.[43] One can only speculate as to whether an unconscious alliance of Cuthbert Gardner and Walter Speed encouraged an impressionable young man to start asking awkward questions about social division and the huge gulf between rich and poor.

Mannock had an early opportunity to discuss the great social issues of the day via a discussion group at the Church Lads' Brigade, and, as he grew older, debates organized by Gardner in his capacity as secretary of St Gregory's Men's Club.[44] This set a pattern for the next five or six years, with Mannock seizing any opportunity to engage in formal debate, even if that meant he himself had to set up the means by which discussion could take place. Prewar social reform and the growth of the Labour movement each rode on the back of a tremendous surge in working-class optimism, self-confidence, and class-consciousness. Compulsory elementary education after 1870, a well-established tradition of autodidacticism among the 'labour aristocracy', and modest gains in free time and disposable income, all enabled the articulate to engage with the 'condition of England' in a variety of forums, by no means all of them visibly political. Chapel, church and meeting-house fulfilled a spiritual and a temporal role, just as they had done since the seventeenth century. Pub and club provided a less puritanical, more informal environment in which to continue the debate, albeit to the obvious disadvantage of women. Mannock had neither the money nor the inclination to wrangle at the bar, and anyway, although the temperance movement was never strong within the Catholic Church, he could scarcely have ignored its influence on his Anglican friends. Nor does he appear to have had much interest in women, his spare time being occupied by the Territorials, a passion for fishing and immersion in the parish life of St Gregory's.[45] If at this time Edward ceased worshipping at his own church then domestic pressure to resume regular attendance at Mass would have been intense. Even in the 1900s it was not uncommon outside Ireland for young unmarried men temporarily to 'lose their faith'. For mothers and priests alike, lapsed Catholics were regrettable, but there was always the comforting thought that marriage and 'a good confession' would ease family tension. But 50 years before the second Vatican Council a poor country woman from Cork would have viewed communion in the Church of England as paramount to supping with the devil.[46] However, Julia's fears that her second son risked excommunication and eternal damnation were eased by the news that Eddie had said farewell to his friends at St Gregory's. Not that she felt any less concern for his future.

Mannock informed his mother that he was leaving Canterbury, even though he would be home every July for the Territorials' annual camp. In May 1911 he started a new life in Northamptonshire, settling in a small but thriving industrial community east of the county town – Wellingborough.

Raising the red flag in the heart of the Midlands

Office life was suffocating, and Mannock had taken the highly unusual step of requesting a transfer from clerical to external duties as a linesman. Such work was harsh and often dangerous. The notion of sacrificing pay and pension in order to rig cable in all weathers was no doubt beyond Patrick's comprehension. Furthermore, his brother could be posted anywhere in England. This of course was exactly what appealed to Edward. In due course he found himself attached to the British Post Office Engineering Department in Northampton, in a crew covering the east of the county. Most days the linesmen had to drive their open-cab lorry to Northampton, where the foreman at the telephone exchange issued the daily roster. While awaiting his orders Mannock would chat to the women on the switchboard, and for a brief while he 'walked out' with one of the telephonists. Interestingly, it was common knowledge within the telephone exchange that 'Paddy', as he was now known, was a socialist.[47] Clearly he didn't keep his views to himself, even at the risk of antagonising his superiors. Even so, Mannock must have exercised some caution. Like most hierarchical organizations in the 1900s the National Telephone Company operated on a basis of deference and respect for authority.[48] Although not as antediluvian and ruthless in its work practices as Rushton & Co., the miserly and corrupt decorators of *The Ragged Trousered Philanthropists*, a Frank Owen-type agitator forever at odds with his foreman, let alone the exchange manager, would scarcely have lasted three weeks, let alone three years.[49]

Riggers were hard men. They needed to be, given the nature of their work. They would have rechristened Mannock on day one – Edwards are rarely found clambering up telegraph poles, perching on thin crosstrees, and reconnecting storm-damaged telephone lines. Here was a man proud of his Irish background, but it must have been his accent that rendered him 'Paddy' to his workmates and 'Pat' to his friends. Ira Jones agreed with MacLanachan that, 'There was a clear-cut incisive timbre in his speech, not exactly Irish but with the definite enunciation and pure vocal sound of the Celt.'[50] Mannock was sending a proportion of his earnings to Julia so he could only afford cheap digs. He lodged with hard working, hard drinking foundrymen in a house belonging to a Mr and Mrs Joyce. The Joyces lived in Elsden Road, in an area known as Eastfield.[51] Squeezed between the station and the town centre, the neighbourhood has changed remarkably little since before the First World War. Take away the cars, cladding and sad 1970s shop fittings, and Eastfield offers an archetypal example of late Victorian urban development. Social zoning was as obvious in Wellingborough as in any other small town faced with an urgent need to accommodate large numbers of unskilled workers and their families. A complex network of terraced rows offered shelter to the wives and children of tanners, brewers, brickmakers and above all boot and shoemakers. Soon Eastfield was home to a new workforce, labouring in the nearby foundries. Wellingborough's traditional industries thrived on the back of the iron works built in 1867 and 1886 to process the ore quarried at nearby Finedon. The blast furnaces operated continuously, lighting up the night sky for miles around and releasing clouds of smoke and ferrous dust. The slate roofs of Eastfield's neat new terraced cottages turned reddish-brown as the dust settled, the environmental price of producing over 100 tons of pig and foundry iron every day.[52]

Industrialization brought in its wake trade unionism and class conflict. Wellingborough had a tradition of radicalism and Dissent. A strong Nonconformist presence was confirmed by the proliferation of chapels, and the make-up of the local Liberal Party.[53] In this respect Wellingborough mirrored the county town, Northampton, in boasting a long tradition of Nonconformity.[54] Eastfield had streets named after Knox, Cromwell and even Gerald Winstanley. Nearby Mill Road, where Mannock eventually made his home, boasted a sizeable Baptist chapel. In the High Street the porticoed, domed and stately Salem Chapel signalled a large and prosperous Congregationalist community, poorer brethren gathering at the Congregational Sunday School off Knox Road. A Liberal Party rooted in manufacturing and Nonconformity fissured whenever trade union solidarity cut across partisan loyalty, and Wellingborough experienced at least three major industrial disputes in the last quarter of the nineteenth century. Unrest arose in the boot and shoe industry, culminating in the spring of 1895 in a six-week lockout across the county: the employers, almost all loyal Liberal supporters, successfully rejected union calls to drop the use of low paid outworkers. However, one issue both sides did agree on was that Tariff Reform would be disastrous for the industry. Staunch free traders, they pointed to the low cost of imported leather and the competitiveness of British shoes in the European market.[55]

If trade unionism was well established in Wellingborough so also was the Co-operative Union, 'the Co-op', selling cheap groceries and meat in Eastfield and, on a much grander scale, in the centre of town. Not that this signalled automatic collaboration between 'Co-operators' and local ILP/LRC activists, however common their interests and natural constituency. If anything, the Co-operative workers allied with the trade unions, notably the railwaymen and the shoemakers, to promote a moderate 'practical' programme of action.[56]

Beyond the strength of the Liberals and the weakness of Tariff Reform, the starkest contrast with Canterbury was the presence of a trades council, a healthy ILP branch, and a Labour Representation Committee (LRC) affiliated to the national body and its parliamentary representation, the 'Labour Party'. Wellingborough, or to be more precise, its industrial quarter, boasted a mature working-class community, with a level of consciousness and a political culture to match. Nor should one assume that Labour in and around Wellingborough focused solely upon Eastfield. Although the Liberals retained a firm grip on Northamptonshire East, Labour was sufficiently well organized that its candidate by no means disgraced himself in the December 1910 election.[57] Because individual membership of the Labour Party was discouraged, or even forbidden, anyone not in an affiliated trade union usually joined via the Fabian Society or the ILP. As there were less than 4000 Fabians in the whole of the country in 1911, and around a third of them lived in London, one can safely suppose that few if any resided in Wellingborough.[58] Thus, beyond the local trades council, membership of the Labour Party meant membership of the Wellingborough ILP.

Long on rhetoric but short on detail, the ILP's 'Big Four' (Keir Hardie, Philip Snowden, Bruce Glasier, Ramsay MacDonald) could still generate enormous energy and optimism as they travelled the length and breadth of the country, rarely flagging in stating their case for a uniquely British brand of socialism.[59] Scarcely pausing to consult either his notes or his Bradshaw, the mellifluous MacDonald could mesmerise a sympathetic audience, glossing over the mediocre performance of the parliamentary party and exposing the inequities of unfettered capitalism. Awash with aims and ambitions, the party secretary rarely paused to explain how all his heady ideals could be achieved, the privileged classes willingly surrendering their power and wealth once universal franchise swept Labour to power. With hindsight, MacDonald is an easy target. Yet highlighting the persuasive power of his rhetoric helps appreciate the degree to which – at a time when Labour carried scant ideological baggage, and no failed record in office – idealism could still triumph over scepticism and disillusion. Because the ILP's socialism had no clear theoretical basis it remained a remarkably inclusive party. Yes, there were splits and breakaways, but overall the level of doctrinal factionalism was far lower than that experienced by fellow members of the Second International. This remained true, even in the final years of peace when the parliamentary leadership's low profile and poor image, undoubtedly forced the ILP across the country to retrench. Ironically, the growth in syndicalism encouraged dissident elements within the ILP to return to the fold, if only to defend their party from the charge of being ineffective because it placed too much faith in a bosses' Parliament.[60] However, for many grassroots activists the politics of Westminster, let alone Hampstead, still remained a mystery. Cynicism was undoubtedly fuelled by the increasingly syndicalist tone of George Lansbury's editorials in the *Daily Herald*, or Fenner Brockway's insistence in the *Labour Leader* that the parliamentary party distance itself further from the government. The flagging circulation of the weekly *Labour Leader* was in itself a sign that by 1914 the ILP's early momentum had waned. The halving of pamphlet sales in two years was another indicator of disappointment with Labour's patchy record since entering Parliament.[61] Nevertheless, for diehard loyalists optimism prevailed, and in this regard Wellingborough was no exception.

A key figure in the Wellingborough ILP was A.E. Eyles, known to everyone as 'Jim'. In 1911 Eyles, who by then was in his late thirties, managed the Highfield Foundry Company, a relatively small business which made mostly pipes and gutters. Eyles lived with his wife Mabel and his young son Derek just down the road from where he worked. Built only nine years previously, 183 Mill Road was no different from the other houses in Coronation Terrace.[62] Eyles, for all his supervisory duties, was clearly indifferent to social status. The men who worked with – and not for – him were neighbours, friends and comrades. Mill Road bisects Eastfield, stretching from the edge of old Wellingborough to the town's first iron works, across the railway line. The 1890s and 1900s saw rapid linear development, and the growth in population was reflected in the size of the Victoria Board School, built in 1895, with both sexes strictly segregated outside lesson time. School-leavers simply crossed the road to commence their working lives in the foundry, the clothiers or the shoe factory.[63] Escape from the humdrum life of shopfloor or scullery came via cheap porter in the Ranelagh Arms, earnest homilies in the Baptist chapel, and gossip in the cornershop. A man could be born, baptized, schooled, employed, pensioned off and buried without ever leaving Mill Road. In the summer of 1911 Jim Eyles invited Paddy Mannock to quit his current digs and move into the next street – he offered him the spare room at 183.

The two men had met at a cricket match in Wellingborough. Eyles casually mentioned to the stranger watching the game that the boil on his neck was so painful it would prevent him from batting. Mannock offered to deputize, but was out for a duck. Conversation resumed in the pavilion, and Eyles was soon seduced by the character and personality of this bright, 'clean-cut young man' who was clearly in need of a new set of tweeds, let alone decent lodgings. Eyles invited him back for tea, and in no time Paddy was one of the family: 'After he moved in, our home was never the same again, our normally quiet life gone forever. It was wonderful really.'[64] Although it is clear from Mannock's letters and Eyles's recollections that they shared an unusually close friendship, it was by no means exclusive. Paddy – now Pat – embraced the whole Eyles family, and both Mabel and Derek duly reciprocated. To apply the predictable cliché, Mannock now felt for the first time in his life the warmth of a close, loving, and above all functional, family. He gave Eyles the impression that this was his first genuine home:

> He came out of his shell because he felt safe with us. He knew that he could air his ideas and not have them thrown back in his face. Some of his concepts were a bit confused due to his rather sketchy self-education, so he really needed to talk them over and put everything into perspective. With all honesty, I can say that Pat developed while he was with us. We gave him the exercise-ground he needed, and it was good to see him grow. He was constantly on the move mentally, his confidence grew, and then he was looking for something else to pose a challenge... The drive for it all came from the years of pain with his family, I am sure of that. One could see the change in his eyes when he spoke of those early years. His hate of it all made him run towards some ideal that only he could see.[65]

Eyles's recollections of Mannock, as quoted by earlier biographers, must be treated with a degree of caution. Understandably, the clear signal is that here was a great man cut down before he had time to come into his own ('He lived as if he had but a few years of life and had to make something out of his time that would make it worthwhile').[66] Naturally Eyles was eager to promote himself as a father figure, but it would be wrong to conclude that henceforth Mannock had little contact with his mother, or with his brothers and sisters: he wrote regularly, and he took every opportunity to visit them. Nevertheless,

Mannock clearly did have a profound impact upon the Eyles family, and vice versa.

With Derek asleep upstairs, his parents and their enthusiastic if ill-tutored lodger would sit until the early hours debating the passage of the Parliament Bill, the growing level of industrial unrest, the suffrage campaign, Loyalist unease and renewed tension over Morocco. Like Lloyd George, Mannock saw no contradiction in abolishing the Lords and retaining the monarchy, or initiating radical reform at home while maintaining the balance of power abroad. Of course his notion of reform went much further than that of the Chancellor, advocating a genuine redistribution of wealth, whether inherited or self-made. Nor was this some idle dream of socialist utopia. For Pat Mannock – like so many Irish nationalists blind to the strength of Unionist resistance – political upheaval signalled revolution not reaction. How he would have relished *The Strange Death of Liberal England* had he lived to read it: 'He told everyone he met that everyman should prepare himself for the new age. The downtrodden of the world were about to get their chance at last; it was a duty for men to make the best of this opportunity...'[67]

Both Jim and Mabel Eyles read widely, and they were soon helping Mannock secure an intellectual underpinning for what up until then had been scarcely more than an inherent sense of injustice. As archetypal Edwardian autodidacts all three saw learning as the key to action, and birth as no bar to the appreciation of high culture.[68] Both of the Eyles were keen musicians, holding a social evening every week where all present were invited to sing or play. They encouraged Mannock to practise on the violin his father had bought him as a child. He worked his way through an assortment of solo arrangements, but still felt happiest playing light classics like 'The Londonderry Air', the tune he played most frequently once stationed in France. Appropriately, his favourite song was 'Danny Boy'. A natural baritone, Mannock had a good voice, ranging from Victorian parlour pieces to light opera.[69] Pat worked his way through Jim Eyles's library, for the first time discovering the Fabians, both Wells and the 'Old Gang' of Shaw and the Webbs. For light reading, in addition to Wells's novels and short stories, Galsworthy and Bennett would have been obvious choices. Keir Hardie and other ILP stalwarts were of course compulsory reading. Still with the ILP, MacDonald's Socialist Library no doubt complemented the myriad of national and branch publications.[70] Mannock's political education lasted throughout his three-year stay with the Eyles family, and 183 Mill Road was his Ruskin or Harlech, his Birkbeck or LSE.

As part of that political education Mannock liked nothing better than a heated argument, but he was also the means by which that argument could come about. In 1908 the YMCA had moved into a large building in Church Street, opposite the town's Post Office. Encouraged by the Eyleses to meet other like-minded young men and women, Mannock became an active member, soon becoming secretary of the YMCA's 'Wellingborough Parliament'. With the Parliament Act removing the Lords' power of veto, and the Liberals' Commons majority dependent on Labour and the Irish, Home Rule was again high on the political agenda. Loyal to his Irish roots, Mannock sat as the Honourable Member for Waterford, the seat of Irish Party leader John Redmond. With a similar sense of occasion, the co-founder of the 'Parliament', one J.T. Johnson, sat for Belfast. Mannock apparently was a good speaker, even-tempered and amusing yet passionate in his beliefs. Even his opponents acknowledged his sincerity, intelligence and commitment.[71] With Jim Eyles so active in the ILP it was scarcely surprising that within a short space of time so also was his protégé. Mannock's organizational skills, already evident in his contribution to the efficiency of Canterbury's Territorial unit, ensured his early election as branch secretary. As such he played a prominent part in the establishment of the Wellingborough Labour Club, the success of which contributed to a rare gain for the party in the 1918 general election, when Walter Smith won the town's newly created seat.[72]

The Wellingborough ILP was expanding at a point when membership of the party nationally had begun to decline. Those activists frustrated by the national leadership's moderation, and its tacit support for the government, were attracted to the breakaway British Socialist Party or even the burgeoning syndicalist movement. Potential new members not in an affiliated trade union now had the option of joining the Labour Party at constituency level, although the LRC's loose federal structure meant that there was still only a small number of local parties on the eve of the First World War. With a growing range of alternatives to joining the ILP, and social discontent manifesting itself increasingly via extra-parliamentary activity, the sharp reduction in the number of branches was perhaps not that surprising. Wellingborough was clearly bucking a trend, at a time when '"active" working-class consciousness in this country was more fragmented and inarticulate than Labour leaders cared to think'.[73] In those ILP branches that were still thriving, the secretary had a crucial role to play, as the party's executive was keen to point out:

> The Secretary of a branch being the centre of its energies, its success or otherwise very largely depends on him. The poorest or weakest

branch may often be made successful by a Secretary who has tact, initiative, and resource, accompanied by good method... the Secretary will usually have to suggest the thing to be done, and the method of doing it...[74]

Eyles insisted that Mannock's 'cheerful countenance and his obvious sincerity' played a big part in getting things done: he rarely alienated people, even when they profoundly disagreed with him.[75]

Eyles could recall only one occasion on which Mannock lost his temper: still a keen angler, he found himself confronted by a water-bailiff insistent that the riverbank was private property. In the true spirit of William Morris, Mannock insisted that God had bequeathed the rivers, lakes and oceans to all mankind, and therefore anybody – rich or poor – should be free to fish unmolested. Luckily he escaped prosecution for either poaching or assault, but the episode clearly rankled.[76] When he wasn't fishing Mannock still relaxed by playing cricket. Joining the YMCA had demonstrated an indifference to Catholic sensitivity already apparent in Canterbury, so it was scarcely surprising that Pat found Methodism the most appropriate avenue for his ecumenical sporting skills. He was first-choice wicket keeper for the Wellingborough Wesleyan Cricket Club. Eyles, the club secretary, also played in the first team, and was considered an exceptionally fine opening bat. Even after he ceased playing in the 1920s he was capable of coming out of retirement to score a century. Unconsciously undermining the 'ace with one eye' myth, Eyles later recalled his team-mate's 'eagle eye and quick thinking'. Similarly, at Leonard Garfirth's house Mannock would show off his hand–eye coordination by landing ring after ring on the highest-scoring hooks of the kitchen ring-board. On Saturday afternoons in winter, Garfirth would invite Paddy home for tea and a game of cards after football. From the autumn of 1911 Mannock played inside right for the Wesleyan Football Club, for whom he also acted as secretary.[77] It is astonishing just how quickly he immersed himself in the life of the community once he moved to Mill Road. Less astonishing perhaps is his absence from the Roman Catholic congregation of Our Lady of the Sacred Heart, the dour, drab church consecrated in 1886 and located less than five minutes walk from Mannock's new home.

Mannock worshipped further along Knox Road at the Neo-Perpendicular church of St Mary the Virgin, designed by the doyen of Anglo-Catholic architects, Sir John Ninian Comper, and dedicated in 1908. Impressive from the outside, the sheer scale of the building is best appreciated from within. The height of the nave generates a sense of awe matched only by one's astonishment on first entering the Anglican cathedral in Liverpool. Unlike Liverpool, one is immediately struck by the light, colour and vibrancy of the nave and choir – a confusion of ecclesiastical styles held together by a Late Gothic rood screen, gilded and carved with the dedication and fine attention to detail so characteristic of the English arts and crafts movement. Even though St Mary the Virgin was not finally completed until 1933, with consecration delayed until 1968 (again, echoes of Liverpool), whenever he entered the building Mannock could not fail to have been amazed by such a triumphant celebration of the greater glory of God. No wonder he strode past Our Lady of the Sacred Heart every Sunday morning when the one true faith could be found, only thinly disguised, in a church that literally soared to the heavens, combining flamboyance and harmony, extravagance and craftsmanship, grandeur and intimacy. Although denied the Marian iconography that came to dominate the church's interwar decoration, Mannock would have noted the 15 stained glass portraits of saints, recusants, and Anglo-Catholics – martyrs and missionaries from the Reformation to the Oxford Movement. Every Sunday Reverend Watts, the first parish priest, said Mass, his liturgy owing little to Cranmer. St Mary's was a high church beacon, and the Peterborough episcopacy's tolerance of its incense and Catholicity no doubt shocked and infuriated the town's evangelicals and chapel-goers.

The generosity of St Mary's patrons, the three Sharman sisters of nearby Elsden Lodge, ensured that the church would embody their Anglo-Catholic beliefs.[78] The sisters appointed the like-minded Comper as their architect, and sought spiritual and material support from the Guild of All Souls, a lay body of wealthy Anglo-Catholics founded in 1873 to restore pre-Reformation practices of burying and praying for the deceased. Equally welcome was the Society of the Holy Cross, founded in 1855, 'to extend the Catholic faith and strengthen the spiritual life of its members'. So also was the Confraternity of the Blessed Sacrament, established in 1863 at a time when Newman and the whole Catholic revival were coming increasingly under attack. Remarkably, all these bodies still exist, their literature freely available in the nave of a church whose worshippers continue to celebrate Mass and Benediction, annually make a pilgrimage to Walsingham, and insist that they, 'belong steadfastly to the one Holy Catholic and Apostolic Church'.[79] Stepping inside St Mary's for the first time Mannock might have been amused to find a ritualism and solemnity appropriate to Westminster Cathedral or Brompton Oratory, but rarely on show in a run-of-the-mill Catholic parish church. Yet he could scarcely have felt out of place once the service began as so much of the liturgy would have been familiar. Not surprisingly he returned the following week, and every Sunday until he set off on his travels in early 1914. Mannock was a bona fide member of the parish of St Mary the Virgin so long as 183 Mill Road remained his registered home address. No wonder, therefore, that beneath the stained glass depiction of St George slaying the dragon, the name 'Edward Mannock' is inscribed in gold leaf on the parish's 1914–18 Roll of Honour.[80]

The inherent conservatism of the Roman Catholic hierarchy under the leadership of Cardinal Vaughan, and the rejection of socialism explicit in the 1891 papal encyclical *Rerum Novarum*, no doubt influenced Mannock's decision not to worship at Our Lady of the Sacred Heart. Blatchford's *Clarion* was not alone in depicting the Vatican as anti-union and indifferent to the worst ills of the capitalist system. Living in the east Midlands, Mannock would almost certainly not have come into contact with the recently founded Catholic Social Guild, nor those Roman Catholic trade unionists who interpreted *Rerum Novarum* as an encouragement to organize (in fact, Leo XIII had insisted that unions had to be approved by the Church). Even a popular and successful figure like John Wheatley could attract fierce criticism from the local clergy when endeavouring to demonstrate that British socialism – as embodied in the ILP – was compatible with Catholic orthodoxy.[81]

St Mary the Virgin ensured Mannock could worship in a familiar environment, but without temporal disapproval of his political beliefs. Many Anglo-Catholics espoused a romantic brand of Christian socialism, 'nostalgic, mediaevalist and ruralist', their simplistic view of the Reformation rendering this second Fall a watershed in the history of mankind, when communalism was succeeded by the cash nexus. This linkage of poverty and Protestantism was encouraged in 1912 by Hilaire Belloc's *The Servile State*, and after the war given at least a degree of intellectual underpinning in R.H. Tawney's *Religion and the Rise of Capitalism*. Tawney, albeit unintentionally, provided Anglo-Catholics with a persuasive case for the Reformation fostering capitalist individualism, and a consequent withdrawal by the Church from its historic role of guiding the faithful through the moral maze of post-feudal economic life.[82] Mannock was far too sensible to be seduced by the notion of an idealized 'Merrie England' offering a model for a post-industrial egalitarian society. Yet he would surely have read Morris's *News From Nowhere* and John Ruskin's highly influential *Unto This Last*. Both of these Victorian sages believed that mechanization had alienated the worker from his or her daily toil, and thus they looked back nostalgically to a pre-industrial –

pre-Reformation – era. Yet the entrepreneurial and almost aggressively anti-church Morris eschewed the naive and essentially reactionary worldview of those Anglo-Catholics who, perhaps with the best of intentions, were set on restoring a pre-Henrician status quo that blithely ignored 400 years of industrial, commercial and colonial expansion. Mannock was attracted, not by medievalist escapism, but by the insistence of Anglo-Catholic socialists that Jesus was a child of the labouring classes crucified for challenging the political and religious establishment of the day. Of course their Nonconformist comrades could share this perception of Christ the divine outlaw, but they viewed with scepticism the now widely shared belief that the sacraments, and most especially the eucharist, facilitated ready identification with, 'the Carpenter of Nazareth, the Divine Revolutionist, the Emancipator of the Oppressed, the Founder of the Democratic Church'.[83] Stewart Headlam's description of the Messiah inspired a succession of mystical and sacramental guilds. Headlam's propagandist Guild of St Matthew was the first band of brothers to tap not only the Catholic revival and 'Parish Communion' movement, but also the 'religion of socialism' which gave the period roughly 1880 to 1895 'its own special dynamism'.[84] The worshippers at St Mary the Virgin – home of the Confraternity of the Blessed Sacrament – placed the eucharist at the heart of the liturgy, with Mannock unique within the congregation in having wholeheartedly embraced the doctrine of transubstantiation the day he took his First Communion. The mystery, and the mysticism, of the blessed sacrament fused with an unambiguously materialist view of the Second Coming: 'the doctrine of the King being secularised, but thereby restored to the pattern of its original exposition by Christ'.[85]

By the time he decided to leave Wellingborough there was little doubt that Pat Mannock could articulate a far more coherent critique of capitalism than when he arrived. Nor was his socialist alternative the naive mishmash of ill-conceived ideas and good intentions that had both charmed and alarmed Jim Eyles when first they met. By 1914, Mannock's idealism was tempered by the intensely practical approach to everything he did. He was after all a skilled telegraph engineer with a natural affinity for machines and an understanding of how they worked. Systems and technology had a natural appeal, and in this respect he could have been a model Fabian technocrat. If roused by the rhetoric of Ramsay MacDonald, Mannock was equally swayed by the Webbs' blue-book version of the New Jerusalem. However, unlike the Fabians, his progressivism was rooted in Christian mysticism rather than humanistic rationalism. Nevertheless, Mannock's success in reconciling Catholicism and ethical socialism necessitated a break with Rome, practically if not doctrinally.

Mannock's enthusiasm for soldiering was by no means incompatible with his religious or political beliefs. Voluntary military service had clear radical associations by the end of the Victorian era, not least within Nonconformist communities excited by the exploits of European nationalists and American abolitionists. Inspired by the martial achievements of Lord Roberts, whose Anglican piety had underpinned his progress from private to C-in-C, 'a remarkable shift from evangelism and nonconformity to military and patriotic allegiance' took place in the early 1900s. Military discipline was now 'powerfully invested with moral and religious qualities': with Anglican militarism fuelling Edwardian notions of insecurity and the need for moral redemption, Territorial drill could comfortably sit alongside High Church worship.[86] At the same time, drawing upon the old Nonconformist association of the Volunteer movement with a distinctly progressive (and continental/ republican) concept of the citizens' militia, by no means all socialists were hostile to the notion of part-time *voluntary* service.[87] Tawney and Attlee, for example, were not alone in dismissing pacifism as utopian, hence their readiness to volunteer in 1914. Both found that military service and shared sacrifice encouraged socialist 'fellowship', and a common purpose conducive to efficiency, industry and meritocratic opportunity. Those beliefs, confirmed by the experience of the trenches, were evident even before the war.[88]

Even those Labour MPs most suspicious of any increase in peacetime military expenditure could see the need for voluntary military service in an era when international relations were becoming ever more volatile. Resolute anti-militarists reconciled their opposition to the government's spending priorities with tacit approval of any proto-people's militia, such as the Territorial Force. The staunchest socialist feared for national security, often sharing the wider public's alarm at the German menace, while at the same time retaining faith – at least in public – in the capacity of international proletarian solidarity to stop the combatant nations in their tracks. British socialists did not have to be as xenophobic as the Social Democratic Federation's H.M. Hyndman in order to denigrate supposedly alien ideas, and by implication those who first conceived them.[89] Both Keir Hardie and Ramsay MacDonald emphasized that their vision of socialism was uniquely British. The formation of the ILP in 1893 signalled the demise of the Marxist – and thus by implication foreign – SDF:

Continental shibboleths and phrases were discarded. The propaganda became British. The history which it used, the modes of thought which it adopted, the political methods which it pursued, the allies which it sought for, were all determined by British conditions.[90]

With sentiments such as these it is scarcely surprising that, 'British socialism came to accept the British state and its specific institutions as the sole legitimate vehicle for the (gradual) advance to socialism.'[91] Paul Ward has argued that this notion of what it meant to be British could easily be absorbed into the dominant national tradition, hence Labour's exaggerated faith in the efficacy of the Westminster system. The mainstream left's faith in the power of government to facilitate fundamental change *ipso facto* excluded those who still refused to see the state as synonymous with the nation. No longer 'radically patriotic', the majority of the left in Britain embraced what Geoffrey Field labelled 'social patriotism...an inwardly focused patriotism, one that is oriented toward social reform and implies some kind of new and improved Britain'.[92] In other words, a reformist Labour movement pledged loyalty to the nation/state, especially in time of national emergency, in exchange for the eventual delivery of secure jobs, better living standards, more houses and an overall improvement in the quality of life for the masses. This of course is exactly how Pat Mannock and Jim Eyles, let alone Sidney Webb and Arthur Henderson, viewed the war – a view seemingly reciprocated by the other side once Lloyd George endorsed the 1915 'Treasury Agreements' as a corporatist quid pro quo between state (and capital) and labour.

Mannock and Eyles – churchgoers, monarchists and civilians in arms – represented the moderate face of Labour, in sharp contrast to the 'godless' Social Democrats in neighbouring Northampton. Their moderation dovetailed with the unashamed reformism of local trade unionists, notably the leaders of the Boot and Shoe Union. In order to reassure voters they played down their 'Socialist' credentials (as indeed did the Northampton SDF in municipal elections). However, this proved counterproductive in that Labour looked little different from the New Liberalism so enthusiastically embraced by a Liberal Party which in Northamptonshire was anything but moribund. Indeed the pro-Labour sympathies of Leo Chiozza Money, the Radical backbencher who sat for Northamptonshire East, had left many trade unionists reluctant to contest the next general election. This left Labour in Wellingborough and the east of the county seriously split, even after the National Executive Committee ruled in 1912 and again in 1914 that a straight fight between Chiozza

Money and the Tories was in nobody's interest. Of course by December 1918 the sensitive issue of whether or not to challenge the Liberals had been overtaken by events: Labour rejected invitations from both Lloyd George and Asquith to ally with their respective wings of the Liberal Party, and chose to fight the 'coupon election' alone. The two new seats of Kettering and Wellingborough were both won by Labour, with large numbers of Liberal voters switching their allegiance (as indeed did their former MP). Locally, as opposed to nationally, Labour was seen as patriotic and pro-war. The largest trade unions in Northamptonshire, notably the ironfounders and the boot and shoemakers, were seen to have rallied to the flag. Party activists were largely pro-war, and known to be so.[93] In addition, the patriotic credentials of the Wellingborough Labour Party were reinforced by the fact that its former secretary was a war hero who – thanks to Jim Eyles – was already acquiring mythical status.

3
Preparing for War, 1914–17

Confinement in Constantinople

A British passport prior to the First World War was a large and splendid document. It appeared the very embodiment of gunboat diplomacy: in large copperplate handwriting Sir Edward Grey, on behalf of His Britannic Majesty, not so much requested as insisted that the holder be able 'to pass freely and without let or hindrance'. Unlike today, it was assumed that border guards from Boulogne to Baghdad would tremble at Carlton Gardens' veiled threat, and with a courteous smile and deferential gesture welcome the effortlessly superior English traveller to their humble nation. In practice, Mannock's passport provided him with precious little in the way of protection once he decided to venture beyond the Bosporos. His photograph depicts a tall, commanding figure, with a taste for fine clothes (for a studio portrait the Sunday suit was still *de rigueur*). With his homburg and walking stick he looks every bit the eligible Edwardian bachelor, smart and suave, a veritable card.[1] But no character of Bennett or Wells ever looked as far as Turkey when setting out to climb the steep ladder of success. Mannock was keen to make his way in the world, and to see as much of it as his limited finances would allow. Seeking one's fortune in the colonies or dominions was relatively commonplace, and it appears Mannock seriously considered mining in South Africa or running a plantation in the West Indies, despite the fact that he was singularly unqualified either to find gold or to grow tobacco. Working in continental Europe, let alone Asia Minor, was almost unheard of, unless one had very specific skills, was unusually adventurous and anticipated ample reward. Mannock spent the winter of 1913–14 deciding where he should go and what he should do, but in the end his mind was made up for him. He heard at work that the *Société Anonyme Ottoman des Téléphones*, the empire's embryonic communications network otherwise known as the Constantinople Telephone Company, had subcontracted its cable-laying operation to the British. The National Telephone Company could not guarantee him a job – in 1914 few managers could envisage transferring even skilled labour from one end of Europe to the other – but clearly there was a good chance of recruitment as a rigger once he had arrived in Constantinople.[2]

Having discussed the idea at length, the Eyles sadly accepted that Pat was determined to leave. Jim and his brother lent Mannock a few pounds, enough for him to take passage on a tramp steamer from Tilbury to Turkey. He walked up Mill Road to the station on 9 February 1914 with the Eyles family insistent that Wellingborough would always be his home, an understanding reinforced by the chilly reception in Canterbury once Patrick heard of his imminent departure.[3] As Mannock's merchantman meandered its way through the Mediterranean a succession of letters to Jim and Mabel confirmed 183 as his first port of call once he resolved to return home, but that, for all the homesickness, he was determined to stay away long enough to ensure he came back a success.[4]

Presumably various ports en route, notably Naples, must have prepared Mannock for the poverty he would encounter once he arrived in the Sultan's capital. Nevertheless, Constantinople must have been a huge culture shock: the noise, the crowds, the oriental atmosphere, and above all the size and splendour of the city. Even allowing for the strong German presence, the western influence in Turkey was still very limited, with expatriots isolated and aloof from the local community. Despite the recent improvement in Anglo-German relations, with agreement on an early completion of the Berlin–Baghdad Railway, most semiofficial newspapers continued to foster hostility towards the British community. Fearful and suspicious, the British tended to keep contact with the indigenous population to a minimum, an option not open to someone like Mannock, whose intention was to secure employment as a supervisor. He would have to work with local linesmen all the time, surmounting the language barrier and overcoming the Turks' deep distrust of anyone hailing from a country once considered an ally and now deemed a potential enemy.

Gladstonian demands for political reform throughout the Ottoman Empire had encouraged nationalist aspirations in south-east Europe, fuelling deep animosity towards Britain among the Turkish military and political elite. Britain's occupation of Egypt in 1882 and growing presence in the Persian Gulf compounded sharp differences between London and Constantinople over the future political composition of the Balkans. Grey's success in 1907 in resolving longstanding territorial differences with Russia was followed a year later by the Young Turk Revolution, both events further worsening Anglo-Ottoman relations. The Foreign Office initiated a rather half-hearted attempt in the spring of 1908 to secure a rapprochement, if only because it was clear that Germany was all too eager to exploit an atmosphere of mutual distrust and antipathy. Mannock's arrival in Turkey coincided with Enver Pasha becoming Minister of War. Enver Pasha had played a key role in the 1908 revolution and was now a powerful member of the reforming Committee of Union and Progress. He was a known Germanophile, having served as military attaché in Berlin for two years, and been instrumental in the appointment in October 1913 of Liman Von Sanders as a marshal in the Ottoman army. By the spring of 1914 Von Sanders was Inspector General, and the German military mission had grown to around 50 officers. Secret talks were initiated on a military alliance, and the deteriorating situation in Europe speeded up the signing of an Ottoman–German Treaty on 2 August, followed by the Sultan authorizing immediate mobilization. The Foreign Office was not unduly surprised by these developments, having largely discounted the upbeat reports of Sir Louis Mallet, the British Ambassador since October 1913. Mallet's credibility in London was undermined by his too long continuing to believe that Enver Pasha and other prominent Young Turks were genuine progressives and not necessarily hostile to British interests. In reality, the British Embassy was almost wholly ignorant of the Committee of Union and Progress's plans and ambitions, and intelligence conveyed to a sceptical Foreign Office was all too often sketchy and speculative.[5]

It was in this febrile atmosphere that, having showed his credentials to the National Telephone Company's contract manager in Constantinople, Mannock was quickly added to the payroll. It appears that his new boss – an engineer from Glasgow, one F. Douglas-Watson – was especially impressed by the notion of setting out for Turkey without the prospect of a definite job.[6] Mannock was taken on as a supervisor, monitoring the progress of various gangs of riggers who were connecting the central telephone service to isolated exchanges out in the countryside. Working in the capital's arid hinterland, in ferocious heat and humidity, he relished the fact that he was no longer a linesman ('No more climbing irons for me!'), but found trying to keep truculent Turkish labourers on task an arduous and often soul-destroying task. The local dialect was almost impenetrable, but he endeavoured to learn something of the language in order to deal directly with problems and grievances. At least Mannock tried to remain even-tempered, unlike most of the

English overseers who were boorish, impatient, intolerant and all too likely to drive their workers away. His subtler approach produced results, and gradually connections with towns and villages out in the countryside were completed.[7]

Success out in the field brought increased responsibility, and with it more time spent on administration and paperwork. When Mannock told his manager that he had successfully applied to the Crown Agents for appointment as an electrical engineer on the Gold Coast he was given promotion as an enticement to stay.[8] Mannock's clerical experience, plus his direct knowledge of laying cable, must have made him an invaluable addition to a harassed British team behind schedule and eager to complete their contract before the international situation deteriorated even further. From his letters to the Eyleses it appears that once he was based almost permanently in Constantinople Mannock began to spend a lot more time socializing with the British community. He must have come across as something of a rough diamond to the more refined members of the embassy and consular staff (he could conceivably have met the newly wed Harold Nicolson and Vita Sackville-West just prior to their departure home – one can only fantasize on that particular encounter). Yet he swam, rowed, played tennis and croquet, resumed horse riding, and would no doubt have gratefully accepted an invitation to play cricket had the opportunity arisen. Although busy organizing the transition to full Turkish control of the new telephone system, Mannock still found plenty of time to sit in the English Club sipping whisky and soda and digesting last month's news from home. He refined his social graces at cocktail parties, meeting 'lots of sweet young English girls out here too, but I am keeping them at arm's length by saying I am married'.[9] Mannock is supposed to have become close to at least one female member of the expat community, sharing a common interest in progressive politics, as opposed to the standard agenda of scandal, social tittle-tattle, and the sad state of the nation (applicable in conversation of course to both Turkey *and* Britain).[10]

This very pleasant existence came to an end in the tense weeks following Britain's declaration of war on 4 August. If Turkey had immediately demonstrated its support for Germany by declaring war on the members of the Triple Entente then the situation would have been much more straightforward: all British nationals would have quickly left the country under the umbrella of the embassy staff's diplomatic immunity. However, power politics at the highest level meant a failure to agree on the most advantageous course of action, and the British government took hasty action to foster those elements within the Porte (court/government) wary of rushing into war. The Foreign Office, assuming an early outcome to the conflict in Europe, endeavoured to keep Turkey neutral for as long as possible. However, across Whitehall the Admiralty signalled its intention to appropriate two expensive new battleships awaiting delivery to the Ottoman navy – an act that further fuelled anti-British sentiment in Constantinople. At the same time, in the eastern Mediterranean a naval engagement on 10 August failed to prevent two German cruisers, the *Goeben* and *Breslau*, from entering the Dardanelles. Enver Pasha bowed to German pressure that the ships be allowed to enter port in order to repair shell damage, and furthermore that chasing British warships would be fired upon if they too sought entry to the Dardanelles. Enver still tried to procrastinate, but now the die was cast. The German crews were given a hero's reception in Constantinople, much to the chagrin of a British community now subject to ceaseless abuse and subordination. The National Telephone Company had assumed that, so long as Turkey remained neutral and the war remote, its employees would be harassed but would remain relatively safe. They would be allowed to get on with their jobs, and their Turkish workers continue to cooperate, simply because it made sense to have the capital's telecommunications system installed and operative prior to commencing military operations. The arrival of the German cruisers speeded up events dramatically, not least because the British Ambassador delivered Grey's demand that the ships be either interned or ordered to leave Turkish waters immediately. Playing for time Enver Pasha, advised by Berlin, informed Mallet that Turkey had entered negotiations to purchase the *Goeben* and the *Breslau*. At the same time, the pro-German faction within the Porte exaggerated traditional fears of pan-Slav expansionism, insisting that in order to survive the Ottoman Empire had to side with one European power bloc or the other, and that the presence of Russia in the Triple Entente ruled out any switch of alliance. A month into the war Sir Louis Mallet acknowledged that Enver Pasha had consolidated control of the administration, and that future action would be necessary to force the Dardanelles. Yet paradoxically the Ambassador somehow still managed to remain optimistic that Turkey would not take the final, fatal step. In London both the Admiralty and the Foreign Office were convinced war was imminent, not least because the Germans were supervising heavy fortification of the Dardanelles and the Turks were inciting the Egyptians to rise up against their colonial masters. Had Whitehall had access to Mannock's letters home then they would have had their worst fears confirmed: 'Things very serious here. War in the air. Great anti-British feeling displayed by the people.'[11]

Convinced by early autumn that there was little prospect of a pan-Islam threat, not least in India, the British began to organize a naval blockade while at the same time withdrawing all nationals employed directly by the Ottoman government, and all consular and embassy staff, including the naval mission. The United States government was formally requested to assume responsibility for all British interests across the whole Ottoman empire should HM Embassy be obliged to close. The State Department acceded to this request, delegating responsibility to Henry Morgenthau, its ambassador in Constantinople between 1913 and 1916.[12] Note that the government did not feel obliged to guarantee immediate evacuation to nationals employed by British companies operating under contract, even though the Foreign Office now rejected Mallet's advice that war would be delayed for several months.[13] Mannock – broke, homesick and increasingly fed up with so little happening at work – spent these final weeks of peace reviewing his prospects. He procrastinated over whether to resign, go home and join up ('I wonder if it will be the end of old Murphy? A soldier's lonely grave.') The delay was to prove fatal.[14]

By the end of September, faced with the closure and mining of the Dardanelles thus sealing off the Black Sea, the Cabinet authorized British agents to foment Arab dissent across the whole of the Middle East and ordered a further military build-up in the Persian Gulf. Aware that Germans were facilitating a major transfer of funds to Constantinople prior to a Turkish naval attack on the Crimea, the British spent the whole of October warning the Porte of the consequences of enemy action in the Black Sea. Odessa and Sebastapol were each shelled on 29 October, and a British steamer sunk at Novorossiysk. Two days later Mallet and his staff left for Athens, the Ambassador still insistent that peace could be maintained (a view shared by the departing Ottoman envoy, similarly discredited at home). On 5 November 1914, following naval action at the mouth of the Dardanelles two days earlier, Britain at last declared war on the Ottoman Empire. Edward Mannock and all other British nationals not given the opportunity to accompany the Ambassador across the Aegean had already been incarcerated for two days, their deportation postponed at the request of Berlin.[15]

According to the American Embassy in Constantinople, most of the British and French residents in the city had been held at the central railway station, where presumably they intended to catch a train to the Greek border. Learning that passengers had been rounded up by troops with fixed bayonets, Morgenthau personally threatened the Grand Vizier that he would 'demand his passports' if the enemy aliens were harmed in any way.[16] At the War Ministry, Enver Pasha indicated to Morgenthau that the delay in releasing the French and the British was purely for administrative reasons, but then warned that, 'if England or Greece attacked any more unfortified towns the Turks' only possible reprisal is to detain all French and English subjects as they cannot send soldiers or ships to these countries'. In other words, Mannock and his companions were being held as hostages, notwithstanding Morgenthau's belief that he could eventually secure safe conduct for all those wishing to leave.[17]

The British responded immediately. Eyre Crowe, long recognized as the most hawkish senior official at the Foreign Office, requested the detention of 'all Ottoman consular representatives in H.M. Dominions'.[18] For the next eight or nine months both sides played the hostage game: the Turks would threaten retaliation against their internees whenever the Allies undertook military action in the Middle East, and in response to such threats Britain or France would take punitive measures against Turkish civilians awaiting exchange. The latter, although small in number, were often prominent figures within the Ottoman elite, which strengthened the Allies' bargaining position. High-level American diplomats acted as intermediaries, constantly resolving stand-offs between the two sides, the complexities of negotiation compounded by the fact that London and France were not always in agreement. The Allies would release one of their more prestigious 'guests' if Constantinople guaranteed better treatment of their hostages, notably not transporting them into the interior or to coastal locations likely to be shelled by the Royal Navy.[19] This latter tactic culminated in May 1915 with a threat to convey all British and French civilians to Gallipoli unless the Allies' naval bombardment ceased. The Turkish justification for taking such action was that civilians were being killed on the peninsula. The British refused to negotiate, insisting that Enver Pasha would be, 'held personally responsible for any civilians transported, hurt, or damaged'. The Foreign Office could afford to take a hard line as the Americans implicitly agreed that leaders of a defeated Ottoman empire could be deemed war criminals on the grounds that they had needlessly endangered non-combatants.[20] Despite the heightened military confrontation, Britain's success in forcing Enver Pasha to back down over sending internees to Gallipoli encouraged greater efforts to swap eminent Ottoman leaders held in Egypt for *all* British and French subjects, and not just the women, children and old men now trickling out of Turkey. The Turks continued to insist that in view of the Dardanelles operation no males between 16 and 60 could leave. Meanwhile the Allies kept capturing more and more Ottoman officials, particularly in Mesopotamia and the

Yemen. Recognizing that their bargaining position was lessening almost by the day, the Turks reduced the age span for male internees to between 17 and 45. Then, in spring 1915, the Americans at last secured agreement on a complete exchange.[21]

British officials were well aware that, despite the Allies' tough bargaining stance, the Turks would not – and did not – hesitate to treat their internees harshly if deemed necessary. Even when the Royal Navy wasn't raiding or shelling Ottoman ships and ports, the conditions endured by Mannock and the other Allied civilians were appalling. Similarly, the treatment meted out by their Turkish guards was little different from, and possibly worse than, that experienced by prisoners of war. Any improvement in food and accommodation was invariably as a consequence of American pressure, with the Embassy's Vice-Consul, Cornelius Van Hemert Engert, directly responsible for the well-being of the British and French internees. Engert was delegated by Morgenthau to undertake the long and tortuous negotiations to secure their release. It would have been his department which reassured an anxious Jim Eyles that, 'Mr Edward Mannock is still in Constantinople and in good health'.[22] A polyglot and former law school lecturer, Engert was a professional diplomat from 1912 to 1945. For the first five years of his career in the State Department Engert held a series of appointments across the Ottoman empire, until at last he himself was briefly interned when the United States entered the war in April 1917. The British civilians were especially lucky to have Engert representing their interests. His private correspondence reveals how poorly he viewed the Turks, being firmly convinced from the time of his arrival in Constantinople that 'this poor race' was fated to lose its empire.[23] However, Engert reserved his deepest hostility for the Germans, and his greatest affection for the British, fighting as they were to defend an innately superior 'Anglo-Saxon civilization', and thereby secure 'the peace and happiness of the whole world'.[24] Thus, the British non-combatants were not dealing with a cold, neutral official, but rather a man who – while always maintaining the diplomatic niceties – was scornful of their captors and hopeful of an early Allied victory. It was the scholarly, genteel, ultra-professional but passionately pro-British Engert who secured the sickly Mannock's repatriation on 1 April 1915, placing him on a train that marked the first stage of a circuitous and physically draining journey across the Balkans. Previous biographers have suggested the internees sailed home from Greece two months after leaving Constantinople, but the journey must have been significantly shorter given that Mannock reported for duty with his old RAMC unit on 22 May.[25]

By the time Engert negotiated his release, Mannock was physically in a bad way, the victim of poor diet and miserly rations, dysentery and recurrent malaria, and a harsh winter spent under canvas and behind barbed wire. All prisoners had suffered similarly, but his poor condition was compounded by punishment beatings, and possibly a period in solitary confinement: he left Turkey emaciated, his body sores incapable of healing without proper treatment and decent food. However, he could not have been as ill as is popularly assumed, otherwise he would never have passed his army medical less than two months later. RAMC doctors in the spring of 1915 would still have been fairly rigorous in determining the fitness of recruits to their own corps, particularly if the individual concerned was a Territorial.[26]

In the early 1930s Ira Jones, with help from the Eyles family, tried to piece together what had happened to Mannock during perhaps the most traumatic four months of his life. While it is tempting for biographers to cultivate the notion that their man was a caged lion, a symbol of resistance and mental toughness, and a role model for the less hardy members of the camp, it is just as easy to dismiss any notion of heroism as mere myth-making. Vernon Smyth and Frederick Oughton conjured up a series of brutal encounters with Turkish guards all too reminiscent of the POW films popular at the time they were writing, the most obvious model being David Lean's *The Bridge Over the River Kwai*. James Dudgeon toned down the violence and melodrama, but still provided a stirring story of Mannock's efforts to smuggle food into the internment camp, and the price he paid for getting caught.[27] Both these accounts are based on Jones's conversations with the Eyleses and correspondence with other prisoners, notably one Florence Minter. It appears that from the outset Mannock regularly goaded the guards, encouraging his companions to demonstrate their keen patriotism, and their apparent indifference to the insults of Turkish soldiers gloating over unsuccessful British offensives in Mesopotamia and Gallipoli:

> I [Florence Minter] see that his comrades invariably spoke of him as a 'sport.' That he was always, and in the dark days of our internment and during that bad time when we were under arrest coming home, he was always cheerful and helpful, and kept the men 'British' all through; he was our philosopher, friend, and guide.[28]

Mannock does seem to have been someone special within a weary and demoralized camp community, even if that final comment clearly has the ring of a judgement made with hindsight: along the lines of 'even then we could all sense that here was a born leader'. Anyone unwilling to bow down and accept his lot was bound to be a target. Jones states that just by organizing a massed rendering of patriotic songs Mannock twice lost the chance of earlier repatriation. However, punishment rations and solitary confinement would be meted out only for genuinely serious offences. Relying on support from his closest aide at the telephone company, Ali Hamid Bey, Mannock appears to have broken out on several occasions, returning with food bought on the black market. Once his nightly excursions were discovered then retribution was swift. A daily diet of black bread and water would have just about destroyed an already weakened frame, even without the fortnight in a concrete box so graphically described by Dudgeon. The box was a fresh development beyond even the imagination of Vernon Smyth, who simply had Mannock spending a few days alone in a tent. Jones made no mention at all of solitary confinement, and it's hard to believe that this was an accidental omission.[29]

Whether or not Mannock was locked up alone, the outcome was still the same: a fierce, almost pathological, hatred of the enemy clearly not evident prior to his incarceration. Indeed, as late as October 1914 Mannock was still wondering whether volunteering for the army was worth 'throwing all my prospects overboard'.[30] Notwithstanding the flowery prose and poor syntax, Jones made an interesting point when he claimed that, 'Mannock's resentment was hardened by the loss of his livelihood at a time when the gates of Life's ambition were opening wide, and at his helpless inaction'.[31] Thus, the caged lion wanted the opportunity to retaliate, but his quarrel with the Germans (the Turks were just a pathetic front) was personal – not only had they invaded Belgium and plunged the whole of Europe into war, but they had stymied the career of a rising young engineer who only a decade earlier had appeared capable of little more than humping potatoes or rinsing shaving bowls. Like all personal quarrels, resentment ran deep – Mannock now wanted revenge.

Agitation in uniform

Once back in England Mannock travelled to Wellingborough to see the Eyles, who found him looking 'an absolute mess'. Jim Eyles fully expected the Army to turn him down, and so he did little to dissuade Pat from rejoining the Territorials. It was clear that Mannock had been deeply affected by his time in Turkey, and that no one – not even his closest friend and confidant – could have tempered his anger and resentment.[32] Returning to Kent, Mannock was able to reassure his sisters and friends in Canterbury that he was at least reasonably fit, and that he had every intention of following his brother's example and joining up. He was unable to see his mother as Julia had moved back to Ireland, living in poor accommodation in Catholic west Belfast, a long way from her native County Cork. Mannock took the train to Ashford, where he found the 3/2nd Home Counties Field Ambulance was at last at full strength, but otherwise had changed remarkably little in the 18 months he had been away. There was very little acknowledgement that a war was being waged just across the Channel, and no suggestion that the unit was in any fit state to go to France. In fact, as Mannock quickly discovered, a surprisingly large number of men fully intended to exercise their right as prewar Territorials not to be posted overseas.[33] The new company commander, Major Chittenden, must have been delighted to regain an experienced transport sergeant, confident in dealing with both horses and men. On 25 May 1915 Mannock, having been given a clean bill of health by the wonderfully named Lieutenant Frank Scroggins, attested to serve a further four years in His Majesty's Territorial Force. The same day he was promoted to acting lance corporal then corporal, and finally sergeant. Exactly two weeks later he became Staff Sergeant Edward Mannock, affectionately known to his fellow NCOs as 'Jerry'. He resumed his former duties, preparing horses, handlers and riders for imminent embarkation.[34]

Mannock was to spend the next nine months preparing for battle, based for most of that time at Halton Park Camp West, near Wendover in Berkshire. Halton Park was the home of Lord Rothschild, who handed the estate over to the Army for use as a huge training, service and support depot. The winter of 1915-16 transformed the neat, beautifully maintained parkland into, in the words of one of Mannock's fellow NCOs, 'a slimy mass of chalky mud, dreary beyond conception'.[35] While Kitchener's New Armies trained ceaselessly for 'the next big push', the medics rolled and re-rolled their bandages, practised wound dressings and enthusiastically joined in mock assaults with no clue as to the horrors awaiting them the following summer at Fricourt or Beaumont-Hamel, Thiepval or la Boisselle.

Mannock was unique within his unit in that he had actually come face to face with the enemy. Only those regulars withdrawn from the BEF to train the new volunteer battalions had come into contact with Germans, and, until the survivors of Gallipoli returned home late in 1915, few soldiers if any had ever seen a Turk. With heavy reliance on the speculative and ill-informed reporting of an ultra-patriotic Fleet Street, and letters from the front heavily censored, most recruits in 1915 were still remarkably ignorant of conditions on the Western Front. For all its physical proximity, the war in France (as opposed to the war in the air, once the Zeppelin raids began) remained a remote phenomenon. The remarkable frankness of films such as *The Battle of the Somme* was still a year away, and with most veterans reluctant to recall in detail their worst experiences of Mons or Le Cateau, the virgin soldier's perception of the enemy relied on atrocity stories and a crude prewar notion of Prussian militarism courtesy of the *Daily Mail* and paranoid fantasists like William Le Queux.[36] It is not surprising therefore that Mannock became increasingly frustrated trying to get across to his comrades what motivated him, and the intensity of his hatred for the Hun. His antipathy was genuine, rooted in direct experience not propaganda. He had quite literally suffered at the hands of the enemy.

Life at Halton Park was mind numbing, but somehow Mannock sustained his fury through hours, days, weeks and months of tedium and monotony. Inaction fuelled anger, not inertia, with Mannock increasingly concerned that his fellow soldiers should have a clear idea as to why they had joined up and what they were fighting for – or, to be more accurate, what they would be fighting for once they finally got across the Channel. Only if officers *and* men had a clear idea of the nation's war aims, and why Germany constituted such a major threat to 'the freedom of civilization', would they train harder and be that much keener to see action.[37] Not only was it crucial to understand what Britain and France (Russia would always be a problem for pro-war radicals and socialists) were fighting *against*, but it was equally important for everyone to have a clear idea as to what the two liberal democracies were fighting *for*. Victory in the field should bring in its wake due reward at home, but only if an enervated working class understood that its wholehearted commitment to securing victory deserved due reward: patriotism could not be taken for granted by the powerful and privileged. At the same time those workers with a key role on the home front had a duty to act responsibly, not out of deference to their employers and political masters but as a demonstration of solidarity with comrades in the trenches. For this reason, a year later Mannock was to be heard vehemently denouncing the strikes on Clydeside. This was despite a sharp rise in ILP membership as a consequence of local party leaders like John Wheatley and David Kirkwood supporting the Shop Stewards Movement (and the rent strikers in the vanguard of the Glasgow housing campaign). Patriotism had to prevail over party loyalty, and indeed Mannock's loyalty to the ILP – as opposed to the Labour Party – must have been sorely stretched by the 1916 conference decision that pacifism was henceforth official policy.[38]

Apathy was anathema to Mannock, and the level of inertia at Halton Park was a microcosm of the 'business as usual' mentality still all too evident 18 months into the war. The Army only began seriously to consider educating Tommy Atkins on war aims, and what the postwar world would be like, in the winter of 1917–18. There was never any real equivalent to the Second World War's Army Bureau of Current Affairs, and yet back in the summer and autumn of 1915 the sergeants' mess at Halton Park pre-empted by a generation the ABCA briefings and discussions that took place in barracks the length and breadth of Britain throughout 1943–4.[39] Mannock, who appears to have been a genuinely popular character and thus less likely to be dismissed as a bloody fool, argued that the mess was an ideal forum for formal debate. Overcoming initial resistance to the notion of openly airing political differences, the idea caught on and, in the true spirit of the Wellingborough YMCA, the weekly debate evolved into a mock parliament. There were over 100 members of the sergeants' mess, and every one of them was entitled to represent a constituency, even if they had never been there or didn't even have a clue as to what the place was like. This time round the 'Honourable Member for Waterford' switched to Newmarket, an obvious seat for such a fine rider but gloriously inappropriate given that most debates contained at least one stormy homily on how every good patriot should also be a socialist.[40] Since rejoining the Army Mannock had never sought to hide his views on the conduct of the war, the ineptitude of the British ruling class, and the opportunity full mobilization offered to sweep aside the old capitalist structure and construct a new order rooted in equality, shared wealth and social justice. Now he articulated these views in front of a large audience, refining his skills as an orator. He forensically exposed the inadequacies of his opponents' arguments, while rarely generating any lasting ill will. In the words of Frederick Oughton, he provided, 'mental food for the men who had, prior to his arrival, been lethargic and indifferent to politics'.[41]

What is interesting about the Halton Park mock parliament is that, although it certainly attracted the attention of senior officers, War Office and RAMC records indicate that no action was taken.[42] Contrast official tolerance of these raucous, noisy, over-crowded and alcohol-fuelled weekly meetings, taking place over several months, with the fate of the similar albeit more ambitious Cairo Forces Parliament set up in the spring of 1944. Although the initiative was encouraged by Captain Gilbert Hall, on behalf of the Army Education Corps, the protagonists were overwhelmingly drawn from the ranks, not least because Labour and Common Wealth dominated proceedings and formed the 'Cabinet'.

For all the claims made postwar regarding the size and significance of the Forces Parliament, regular attendance was not that much greater than the numbers crammed into the sergeants' mess nearly 30 years earlier to hear Mannock in full flow. Thus, although around 500 met in the Cairo Concert Hall on 5 March 1944 to hear Labour deliver its 'King's Speech', only 229 had actually voted in the preceding mock election. Nevertheless, the very existence of such an assembly was sufficiently serious for Middle East Command to take swift action following questions to the Secretary for War in the Commons: all press coverage was censored, future Labour MP Leo Abse and other high-profile speakers were posted home, and a War Office inquiry recommended that Hall resign his commission immediately.[43]

Apart from the fact that knowledge of the parliament was restricted to the camp, the main reason for an absence of any similarly draconian intervention at Halton Park may be that debates were restricted to senior NCOs, many of them non-combatants, and therefore the risk of dissent and outright insubordination was low. It took the introduction of conscription in 1916, and revolution in Russia a year later, to galvanize concern in the War Office as to the reliability of either battle-weary volunteers or resentful conscripts. Even then, any protest was *ad hoc* and focused upon specific grievances – witness the demands of the Workers and Soldiers Councils in the summer of 1917. Maybe its authors remained unrepentant in their commitment to destroy the capitalist edifice, but the twelve-point petition actually adopted at Tunbridge Wells in June 1917 contained no hint of revolution or subversion, its moderate and reformist demands representing, 'pre-eminently the voice of the respectable working man, the sort of chap who has traditionally formed the backbone of the Labour movement', for which read 'Jerry' Mannock.[44] Thus another reason for official tolerance of the mock parliament was almost certainly the tone and content of the speeches. Speakers supported the war effort, and were not calling for a negotiated peace. Nor were they attacking senior staff officers such as Sir John French, in 1915 still in command of the BEF. Any criticism of the conduct of the war focused upon the politicians, at a time when most senior members of the first coalition remained Liberals.[45] Only a year into the war the officer corps had yet to broaden sufficiently in social composition so as seriously to undermine peacetime Unionist sympathies. While no doubt horrified by Mannock's socialist vision, any snooping officers would have relished his ferocious denunciation of Asquith's leadership and Lloyd George's mendacity, and been heartened by his patriotic exhortation to, 'fight to the last man! We must kill every enemy.'[46]

Mannock's company commander would have concluded that here was someone who might have some rum ideas, but was nevertheless exactly the sort of chap you wanted to follow you over the top. If there was any query concerning Mannock, it would have been why he was in the RAMC and not serving in a front-line cavalry regiment.

This was a question Mannock was increasingly asking himself, albeit without any suggestion of riding into battle. If he found himself in a combat role he wanted to be able to draw upon his engineering experience and expertise. Mannock found it increasingly difficult to reconcile his view of the enemy with his obligation, in theory at least, to treat friend and foe alike. He felt frustrated by the prospect of going to war but being asked to act in a wholly humanitarian fashion, however laudable and demanding such a role might be. It was increasingly obvious that he should transfer to a combatant unit, and he was forced to make a decision when news came through that the RAMC's transport personnel would be transferred to the Army Service Corps prior to posting overseas. Technically, soldiers in the ASC had combatant status, but once in France it would be that much more difficult to transfer out. Mannock had to move quickly, and in November 1915 he applied for a commission, with the full support of his commanding officer (further confirmation of the authorities' relaxed view of the heated discussions taking place in the sergeants' mess).[47]

Mannock sought appointment as an officer cadet in the Royal Engineers, and his application was strong: not only did he have a fine service record to date, but he could list Douglas-Watson and also Stewart Campbell, the chartered secretary of the Wellingborough Technical Institute, as referees willing to confirm his ability as an engineer. Ironically, although ideally qualified to join the BEF's huge operation of laying literally thousands of miles of telephone wire across Belgium and northern France, Mannock intended to train as a tunnelling officer, one of the dirtiest and deadliest jobs on the Western Front. He informed Jim Eyles that he was going to, 'blow the bastards up. The higher they go and the more pieces that come down, the better!'[48]

War Office bureaucracy meant Mannock did not actually leave Halton Park until early 1916. He was transferred to the Royal Engineers Signal Section at Fenny Stratford in Bedfordshire, convenient for a 48-hour leave in Wellingborough should the opportunity arise. Until commissioned, he was technically still a Territorial, and even when gazetted second lieutenant he remained on probation. Mannock in fact became a 'temporary gentleman' in the Royal Engineers on 1 April 1916, and a full second lieutenant the following June. A senior NCO and in civilian life an engineer, Mannock was typical of a new generation commissioned in the corps and less fashionable regiments once casualties mounted: new weapons and support systems demanded men with organizational and mechanical skills irrespective of their social background. By the final year of the war no less than 40 per cent of British officers were of working-class or at least lower middle-class origin.[49]

Initial training lasted three months, and Mannock quickly found life in the Royal Engineers a huge disappointment. The tedium of service life was no different from Halton Park, with short bursts of intense activity interrupting long periods when there was little to do but read, smoke and engage in desultory conversation. Mannock avoided falling into the trap of drinking heavily, if only because he was old enough to appreciate that he should do nothing to create a poor impression in the eyes of his instructors.[50] Like most soldiers, he smoked heavily, but usually a pipe. The pipe-smoking, cautious, intense and well-organized veteran of nearly a decade's military service off and on – a man who, even if he was reluctant to talk about it, had clearly been around a bit – could easily have found that his age was a positive advantage. Naive young graduates and ex-sixthformers could conceivably have learnt a great deal from an old sweat whose knowledge of the Army stretched far beyond the basic drill, obsolete weaponry and khaki wide games of the OTC. However, the arrogance of youth fused with the insufferance of the Edwardian upper middle class – the notion of learning anything from a man nearly 30 who spoke with a strange accent was deemed risible. Furthermore, Mannock never sought to hide his working-class roots: he felt no shame in who he was and where he came from. Similarly, he took pride in his political beliefs, rarely exercising discretion if ever discussion drifted into matters graver than girls and gaiety.[51] It was not just a case of Mannock being alienated from most of his fellow cadets by dint of age and class (and of course religion, as would have become clear on the first Sunday in camp if he opted not to worship as 'C of E'). What was equally obvious was that he adhered to a fundamentally different set of values from those of his new comrades, their simple, straightforward view of the war still moulded by the chaplain's sermon and the headmaster's speech, the tutor's valediction and *The Times*' benediction.

Notwithstanding croquet and cocktails in Constantinople, this was the first occasion that Mannock was in close proximity for an extended period with young men from very different backgrounds, with whom he had precious little in common other than cricket and horses. Thus he rarely felt relaxed, and the result was that he came across as both aloof and aggressive. The jollity and junior common room chit-chat irritated him intensely. Mannock unfairly saw his fellow cadets as languid drones, with little serious interest in the war other than how it directly impacted upon them. Too often his annoyance manifested itself in tirades against the English class system, which served little purpose other than to make him even more unpopular. Prejudice fuelled prejudice, and the poverty of Mannock's argument merely reinforced his listeners' stereotyped perception of the ill-educated labouring classes. Mannock was much more impressive when not roused to anger. Then, he could marshal his case in an orderly and systematic fashion, as fostered by Jim Eyles before the war. While clearly influenced by the likes of Wells, Shaw and Hardie, most of Mannock's ideas were firmly rooted in personal experience, the assistant adjutant at Fenny Stratford insisting, 'His opinions were not second-hand, but came clean and fresh-minted from a keen and original brain.'[52] As always, we must be cautious of those who knew him not over-egging the cake. On the other hand, Lieutenant Buchanan was not the first, nor by any means the last, to note that Mannock 'hated shams of every kind. Snobbery made him writhe with contempt. Cant, hypocrisy and pretence he loathed...'[53]

One event that occurred while Mannock was in training at Fenny Stratford was the Easter Rising in Dublin. A Home Ruler by instinct, Mannock's position at the start of the war is best summed up by a fellow officer much closer to the republican fault line, the Dublin poet Francis Ledwidge:

> I joined the British Army because she stood between Ireland and an enemy common to our civilization and I would not have her say that she defended us while we did nothing at home but pass resolutions.[54]

On leave when the Rising broke out on 24 April 1916, Ledwidge chose not to desert, dying in Flanders a year later. Mannock experienced no such conflicting loyalties. He was opposed to militant republicanism at the best of times, so the notion of exploiting Britain's present difficulties would have been doubly abhorrent. In this respect he was no different from the majority of Irish soldiers on the Western Front when they first heard news of the fighting in Dublin. He might have been uneasy at the subsequent treatment of the rebels, but like most officers his moderate brand of nationalism would soon have 'felt out of place in the new political culture of Sinn Fein'. Nevertheless, his hope was that 'England may keep faith / For all that is done and said.'[55] We can be fairly sure how 'Murphy' felt about the Rising, but not of course certain. Nevertheless, the violence in Ireland warrants mention as it must surely have heightened tension between Mannock and the other cadets, if only because there would have been a torrent of disparaging, and no doubt racist, comments about the Irish.

The easy-going Buchanan took a more generous view than Mannock of the 'high-spirited, cheerful, happy-go-lucky crowd' that made up the rest of RE Class S.6. At the same time, however, he came to appreciate just how much Mannock was focused on the war. Impatient for a posting, the senior member of the class found much of the basic training repetitive and irrelevant, all too often covering familiar ground.[56] He knew King's Regulations inside out, had drilled for a decade, and could ride like a hussar. Not only that, but the Wellingborough Technical Institute and the National Telephone Company had given him a solid grounding in basic science. Mannock grew ever more impatient as pedantic inspectors and turgid army textbooks endeavoured to fill the gaping holes in these expensively educated cadets' knowledge of elementary physics. Fed up with the routine, and even more fed up with his companions, all he wanted to do was to get to the front before it was all over.[57]

As assistant adjutant, Buchanan had immediate responsibility for Mannock's progress, having recommended his acceptance on the training course in the first place. Buchanan's recollection of his first meeting with Mannock offers a rare portrait of the proto-action man a year after his return from Turkey. He looked like a uniformed Richard Hannay, and indeed Buchanan's prose style is reminiscent of his near namesake. ('A tall, hard-bitten-looking fellow stood before me, with more the appearance of a Colonial than an Englishman, blue-grey eyes, a thin clean-shaven face, and a rather grim impression.') Mannock's uniform was immaculate, but unfortunately his riding boots were being repaired, so he was forced to strap his spurs over a pair of wellingtons.[58] This bathetic scene can be read in two ways. On the one hand, here was a man so single-minded and self-confident that he simply did not care what others thought of him, and on the other, this was someone so naive and gauche that he scarcely appreciated the absurdity of his appearance. The former seems the more likely, although at this stage in Mannock's military career there remains an attractive innocence, which would explain how his pride prevented him from borrowing a pair of boots – an obvious course of action familiar to anyone who had ever boarded at school or weekended in the country. It is scarcely surprising, therefore, that Buchanan was astonished to discover one of his most cultured and sophisticated friends, Frank Mannock, was related to this awkward and uncomfortable, somewhat lugubrious, candidate for the King's commission.[59]

Buchanan had an obvious motive for taking an interest in Mannock's progress, and he clearly believed that he forged some sort of a relationship with him as Class S.6 painfully progressed to completion of the course. Friendship with the assistant adjutant must have been a key factor in Mannock successfully securing a transfer to the Royal Flying Corps not long after his commission was confirmed in June 1916. Having already switched from the RAMC to the Royal Engineers, trying to convince a sceptical CO that at the age of 28 he now wished to train as a pilot necessitated strong nerves and enormous self-motivation. Mannock would have needed all the help he could get, including a quiet word from Buchanan in the colonel's ear. To begin with Buchanan urged Mannock to think again, considering him too old and thus incapable of adapting to a strange and risky form of warfare best suited to young men with few fears and fast reactions.[60] In the end, however, Mannock's latest mentor acquiesced, and helped him survive a stormy interview, which ended with a formal recommendation of transfer to the RFC. With pilots at a premium the War Office responded speedily, and Mannock took a medical board at Bedford Military Hospital on 2 August 1916. This of course is the key moment in the story of 'the ace with one eye' – the occasion on which he supposedly duped the doctor by learning all the letters on the Snellen chart while the orderly was out of the room.[61] As pointed out in Chapter 2, Mannock's excellent test results were exactly the same as those recorded twice in the previous year, and indeed, when examined in November 1915 he had declared 'upon his honour' that he suffered no impaired vision, and had withheld no information regarding his current state of health. Not surprisingly, therefore, despite labelling him 'circulatory, nervous', the board passed 2/Lieut. E. Mannock RE as fully fit, and ready for immediate entry in to the Royal Flying Corps.[62]

How then did Mannock arrive at this decision to train as a pilot, an undertaking for which at first sight he appeared particularly unqualified and unsuited? Firstly, there was no reason at all why he could not learn how to fly in combat. Buchanan clearly thought he was too old, and that flying was a young man's game, but then he mistakenly presumed Mannock to be over 30.[63] A fit male in his late twenties would not noticeably have slowed down in terms of speed of response. Admittedly Mannock knew little about aeroplanes, but he did have a basic grounding in physics and mathematics at a time when the science of aerodynamics was still in its infancy. Similarly, he had an affinity for machinery, and a ready capacity to learn – and learn quickly. Also, despite his spell in the RAMC, he was comfortable handling guns, and at Fenny Stratford must surely have been introduced to automatic weapons. So, although Mannock did not fit the social/age profile of the typical RFC pilot, he was by no means an exception to the rule, particularly as more and more non-commissioned observers progressed to officer/pilot training. He was as well qualified as any other serving officer attracted to the notion of waging war in the air.

Secondly, Mannock had been appalled to learn that training to be a tunnelling officer was a lengthy process, that could take anything up to a year.[64] Quite simply, he could not face the prospect of his arrival in France being delayed a further 12 months, particularly given the false optimism that was so all-pervasive in the final weeks of preparation for the Somme. What if Haig was right, and this great offensive really did break through the German line? Mannock had as yet to see what conditions were really like on the Western Front, so the notion of British troops securing a breach in the enemy's defences through which the cavalry could pour did not seem that far-fetched. What if the war really was over before the brass hats deemed him ready to lay mines deep under the enemy trenches? Mannock concluded that he could not afford to wait any longer – if the Royal Engineers could not provide a speedy passage to the front line then he simply had to look elsewhere.

Thirdly, Mannock became a pilot because in his ignorance he saw it as a glamorous alternative to life in the trenches. In this respect he was no different from thousands of other young men attracted to the RFC and the Royal Naval Air Service (RNAS). There is no evidence that, before passing out and going on leave in June 1916, he was contemplating a second transfer. After all, if he definitely did not want to be a tunnelling officer then his knowledge of telephones would have earned him a place on the next boat train to Amiens. Vernon Smyth and Frederick Oughton claimed that Mannock was first attracted to flying during a conversation with an old workmate from Northampton, Eric Tompkins, whom he met on the platform at Bedford railway station while waiting for a train to Wellingborough.[65] Tompkins was serving with the RFC in Wiltshire, at the Central Flying School (CFS), Netheravon. Despite its name the CFS was no longer the Royal Flying Corps' only training centre. Reserve Squadrons of the newly formed, and rapidly expanding, Training Brigade were based at a number of airfields scattered around the British Isles.[66] Tompkins painted an attractive picture of life on and above Salisbury Plain, but he didn't hide the fact that as many if not more pilots died in training as in combat.[67] He may have tried to temper his enthusiasm for aircraft, but by the time the two men went their separate ways he had captured Mannock's imagination. It would not be an exaggeration to say that the 30 minutes spent with Eric Tompkins in the bar of the Bedford

Station Hotel marked a watershed in 2/Lieut. Mannock's so far uneventful military career.

Tompkins wasn't alone in portraying the skies above France as a jousting ground for the chivalrous knights of the air. By the summer of 1916 British newspapers were beginning to discover Albert Ball, who was still only 19 when awarded the Military Cross on 1 July.[68] Ball's first posting to France saw him credited with 30 'kills' in just over seven months, adding a DSO and two bars to his MC. His letters home record the close personal interest shown in him by senior staff like Hugh Trenchard, the RFC's commanding officer in France and in 1918 the first Chief of the Air Staff. Ball was in many respects the RFC's first proper air ace. His exploits helped raise the profile of what was still a fledgling service, and yet Trenchard strongly objected to the term 'ace' on grounds of morale. Neither he nor other field commanders liked the term, arguing that the contribution of ordinary service personnel would be ignored once attention focused on individual pilots. This attitude contrasted sharply with the Germans and the French, who were the first to label pilots like Garros and Pègoud aces. Both countries recognized the propaganda value of their fighter pilots who, by 1916, had caught the public imagination in Paris and Berlin. Although initially a *jagdflier* had to score ten 'kills' to become an ace, eventually the German Air Service came into line with the Allies, and five became the accepted score. As the rate of attrition grew, and the killing potential of experienced pilots and superior machinery increased, so a top pilot became for all intents and purposes an ace of aces. Ball demonstrated that, despite the best efforts of Trenchard and others to play down individual achievement, newspapers would sensationalize the war in the air, and give their readers heroes resonant of an earlier, cleaner, more honourable age. Ironically, for all the attention given to Ball as a clean, uncorrupted, almost saintly young man, quite literally rising above the dirt and carnage of the trenches, this new fascination with 'kills' was macabre, albeit never as obsessive as in the French and German press.[69] Ball was both the first and the last British pilot to secure a place for himself in the wartime pantheon of national heroes. Although McCudden attracted publicity in early 1918, as did the Canadian ace 'Billy' Bishop, Mannock's death attracted only modest newspaper coverage, and the same was true for other British aces. Recognition was largely postwar, and in many cases posthumous. Thus only a small number of British pilots became household names during the war, unlike in France and most especially Germany, where members of Richthofen's Circus would feature on postcards widely available in Berlin and the other big cities of the Reich.[70] In Britain only Albert Ball achieved the level of recognition of an Oswald Boelcke or a Lothar von Richthofen, let alone the Red Baron himself. However, Ball left a permanent imprint upon the national psyche: over 80 years after his death he is invariably the one British pilot anyone questioned about the First World War can actually put a name to.

A loner like Bishop, Ball played a key role in the air superiority secured by the RFC during the early stages of the Somme offensive in July 1916. His final posting was with the elite 56 Squadron, testing the new SE5 during 'Bloody April' 1917, when outgunned aircraft and insufficient pilots left the British reeling from their worst losses of the war. By the time he was shot down, near Lens on 7 May 1917, Ball had destroyed 43 aeroplanes and one balloon. He bridged the transition in tactics and technology from the early dogfights, where scout pilots frantically sought the best position from which to fire their wing-mounted machine guns, to the formation flying of 1917–18, where their successors endeavoured to gain the edge in speed, manoeuvrability and weaponry.[71] Ball's service in France just overlapped with Mannock. As if passing on the baton, Mannock secured his first success – downing a balloon – on the day Ball died.[72] Both men were in character very different, not least because of their respective ages and social background – Ball's father was a property dealer and prominent local politician, serving as Mayor of Nottingham twice before the war and twice afterwards. Also, Mannock was far less of a solitary warrior than Ball, and yet both men shared a similar approach to aerial combat: neither refrained from telling Trenchard to his face that the SE5 was not up to the job; neither stopped flying when it was becoming ever more obvious that they were cracking up; neither were prepared to put their faith in the squadron's armourers when it came to sighting and loading their guns; neither lacked the technical expertise to fine-tune their aircraft, particularly the engines, to maximize performance at high altitude; and neither were reluctant to pass on to their comrades the benefit of their experience.[73] In addition of course, both men were to be awarded the Victoria Cross posthumously.

Ball's new found fame fired Mannock's attention, and while staying with the Eyleses on leave he found himself scouring the papers for reports of the war in the air. When, later that summer, he received confirmation of his transfer to the RFC, he recorded his 'unbounded delight' in his diary:

> Now for the Bosche! I am going to strive to become a scout pilot like Ball. Watch me. I wonder what fate has in store?[74]

Planes before politics – learning to fly

The years 1916–17 marked a major expansion in the training programme for pilots, both in Britain and in the Middle East. Not only did the RFC and the RNAS need more personnel, but each of the services wanted pilots with the skills and flying time to ensure that, if they did die soon after their arrival at the front, then it was due to enemy action and not their own inexperience. More stringent qualification tests were introduced, with trainee pilots progressing from one level to another, and a minimum set in December 1916 of between 20 and 28 hours solo flying prior to posting. The training programme was reformed and restructured, but in the second half of 1916 there remained a very real problem of too many trainees and not enough aircraft or instructors. Nor was there any consistency in the quality of instruction, with no formal training of the trainers. Not surprisingly, experienced pilots found going up in obsolete, poorly maintained aircraft, and at the same time placing their lives in the hands of a novice in the front cockpit, a nerve-wracking and highly dangerous activity. Too often instructors were either veterans scarred by spending too long in the front line before being sent home to recover, or they were pilots who for various reasons were not deemed up to the job while serving in a front-line squadron. There were also of course some exceptionally able instructors, but their contribution was too often overshadowed by the insensitive and cynical methods of their mediocre colleagues.[75]

Mannock only came into contact with these stressed-out fliers after a six-week induction course at the RFC's No. 1 School of Military Aeronautics, Reading.[76] The course was almost all theory, covering everything from navigation to aerodynamics, but there was the chance to go up in a training aircraft, such as the relatively sophisticated Avro 504. Mannock, who now for the first time found himself addressed by all and sundry as 'Mick', passed out with honours, an early indicator of how comfortable he would be with the mechanics of flying. Throughout October and November 1916 he was based at Hendon, being taught how to fly. Even two years into the war pilots had to secure a 'ticket', the flying certificate awarded after passing a test administered by the Royal Aero Club. It was still possible for an impatient volunteer to secure the 'ticket' privately if he found that all RFC courses were full up and he could afford to pay for civilian instruction. Thus Gwilym Lewis, who flew with Mannock in 40 Squadron in the winter of 1917–18, paid for his own training at Hendon in order to ensure his transfer from a Territorial infantry regiment to the RFC.[77] Mannock gained his 'ticket' – Royal

Aero Club Certificate 3895 – on 28 November 1916, completing his induction to the RFC at 19 Training Squadron, based at Hounslow in Middlesex. On 1 February 1917 he became a probationary flying officer, having met the increasingly stringent criteria established by the War Office in the course of 1916.[78]

While at Hounslow Mannock was sent to 28 Reserve Squadron at Castle Bromwich, where he was stationed over the Christmas period. Legend has it that, while flying over the east Midlands, his engine cut out and he chose to land on the playing field of Wellingborough School, for Mannock one of the town's most visible symbols of wealth and privilege. Mannock phoned for help, and was told that a Crossley tender would bring two mechanics. In the meantime he strolled round to Mill Road in order to surprise the Eyleses. They insisted that he stay for Christmas dinner, but Pat declined as he had to get back to the school in order to meet the ground crew. The fitters arrived, only to taxi the aircraft into a tree while testing the engine. More help was needed, so Mannock now had a clear conscience – he went back to eat with the Eyleses, leaving the mechanics eating their dinner courtesy of the junior school headmaster.[79] This reads like an urban myth, particularly the idea that Mannock should be flying on Christmas Day, which, when on active service a year later, he absolutely refused to do.[80] Also, it seems hard to believe that such an inexperienced pilot should be flying solo in poor weather and so far from home. Assuming the incident did occur – and elderly people in Wellingborough as late as the 1980s were insistent that Mannock had once landed at the school – was it necessarily on Christmas Day?[81] Why was G.H. Oxland, the master who provided the mechanics with a meal, able to do so? What was he doing there? In other words, was the story of having Christmas dinner a seasonal embellishment? If Pat Mannock really did visit Jim and Mabel Eyles on 25 December 1916 then an engine failure over Northamptonshire was remarkably convenient. It would have been too much of a coincidence – would Mannock have risked court martial and dismissal from the RFC in order to pull off such an outrageous stunt? At some point later in the war, perhaps January 1918 when he returned home from serving with 40 Squadron, Mannock probably did fly into Wellingborough, but not to eat roast turkey and plum pudding. That story has become part of urban folklore because after all it is a good one, fuelling the Mannock legend – he bucked authority, even to the extent of going home for his Christmas dinner.

After leaving Hounslow, Mannock spent the next fortnight on the Kent coast, at the School of Aerial Gunnery. In Hythe he underwent intensive instruction in use of the Lewis machine gun, but still found time to visit friends in Canterbury, only a short train ride away. He was now at the stage where he needed to fly every day in order to build up the number of hours recorded in his log book, to gain in self-confidence and, most important of all, to become relaxed when handling his aircraft.[82] Mannock felt at one with his machine from the first time he sat in the cockpit, his instructor at Hendon noting the ease with which he took to flying. There were obvious technical shortcomings, not least an inability to master the art of the smooth landing, but when in the air Mannock was wholly uninhibited. He had a light touch on the controls, and he appeared genuinely to enjoy himself.[83] He was by no means a natural pilot, like Ball or McCudden, but he had an enviable capacity to switch off and relish the simple pleasure of flying, even when the fighting was at its worse. In an unusually lyrical passage in *Fighter Pilot* MacLanachan described Mannock in a 'world of glittering white iciness' oblivious to the war below the cloud bank, and 'contour chasing among the steep feathery mountains and valleys' in an atmosphere of complete calm: 'It was indeed an entrancing world in which our S.E.s were suspended, and very soon the rest of us realised that it had cast its spell over Mick'.[84]

By this time, although still irritated by anyone who seemingly did not share his intense feelings about the war, Mannock no longer felt the need to initiate debate in the same way as he had at Halton Park, and probably tried to do at Fenny Stratford. He had a poor opinion of senior air staff at Hendon, and this was reciprocated. Wrongly, they judged him opinionated, and too often ignorant of polite behaviour and mess convention. With training aircraft in short supply, and no spare capacity if one crashed, Mannock was severely reprimanded for making an unauthorized flight in the primitive and dangerous Caudron Biplane. A fanciful story in *Ace With One Eye* suggested Mannock's card was marked the day his colonel spotted him arm in arm with an attractive canteen worker deemed unfit for an officer and a gentleman. Even in the mess of RFC Hendon, Mannock continued to wage the class war, indifferent to the bluster and threats of the station commander.[85] Whether he did or did not clash with his CO, Mannock clearly impressed his instructor at Hendon. The same was to prove true of the final stage in his training to become a scout pilot. Posted to 10 Reserve Squadron at Joyce Green, near Dartford at the mouth of the Thames, Mannock convinced his instructors that he had the right temperament to fight in France. Unfortunately, his commanding officer, a temperamental major from South Africa called Swart, suspected exactly the opposite.

Mannock spent around six weeks at Joyce Green, a bleak airfield amid the marshes of the Thames estuary which doubled up as an advanced training school and a base for the defence of London against Zeppelin attack. On the training side morale was poor, with the novice pilots deeply resentful of the scarcely veiled hostility shown towards them by several of the instructors. While the latter had seen little or no service overseas, many of those they were training had served for long periods on the Western Front before transferring to the RFC.[86] Early in March 1917, not long after Mannock had arrived at Joyce Green, a very different sort of tutor arrived, albeit only for a short stay. Lieutenant James McCudden spent the spring of 1917 in England as an instructor, and later on as a night fighter once the giant Gotha bombers began raiding London. Holding both the MC and the Military Medal, a reminder of his rise from the ranks, McCudden's reputation was second only to Ball. Indeed, when McCudden was posted as a flight commander to 56 Squadron in August 1917 he was perhaps the only pilot capable of expecting immediate respect from Ball's former comrades. Mannock and McCudden met for the first time at Joyce Green, and immediately hit it off. Although McCudden was already sufficiently famous as to be recognized at the Gaiety and the Savoy, and Mannock had yet to fire a gun in anger, they recognized how much they had in common. Seven years younger than Mannock, McCudden was born into a large Irish Catholic family, his father an NCO in the Royal Engineers. Being brought up in a service family in the Medway towns, 'Jimmy' McCudden's background was remarkably similar to his new friend, the big difference being the powerful presence of his father and his enlistment as a Regular at the earliest opportunity. In the ten years he had been in the Army McCudden had risen from boy bugler to lieutenant, and later in the year he would be promoted to captain. Having transferred from his father's corps to the newly established RFC in April 1913, he had progressed from engine fitter to flying officer, via a brief period in 1915 as an observer/gunner. McCudden's year flying DH2 pusher scouts with 29 Squadron had established his reputation as a skilled pilot and marksman, earning him the MC once his tour of duty ended in February 1917.[87]

Although awaiting a new Bristol Scout in which to show off his skills, at Joyce Green McCudden had to rely on the ageing but reliable DH2 to demonstrate basic airfighting manoeuvres. Only one of Mannock's class felt ready to emulate their guest instructor's tricks, especially when he 'spun' his aircraft. Mannock questioned McCudden closely about how to put the DH2 into a spin, and most especially how to pull out of it. McCudden's first impression of Mannock was that here was a, 'typical example of the impetuous young Irishman, and I always thought he was of the type to do or die'.[88] This judgement was exactly right, because the day after McCudden insisted 2000 feet was the minimum height from which to go into a spin and survive, Mannock made his first solo flight in a DH2. Although he later made out that he got into a spin by mistake – an explanation McCudden, and later Ira Jones, were happy to accept – Mannock's more recent biographers have echoed the view of Major Swart. The squadron commander was absolutely convinced his tyro pilot's spin was no accident, but a deliberate attempt to impress McCudden.[89] Oughton and Smyth depicted a naive Mick Mannock determined to demonstrate that the DH2 was not the death trap most of the other students believed it to be. A further incentive was to prove that even an inexperienced pilot could survive a spin from below 2000 feet. In a rare passage of fine writing Oughton's technical expertise and Smyth's keen sense of drama combined to give the lay reader a clear idea of how Mannock dropped from 1500 feet, stalled his engine, brought it back to life, flicked the aircraft into a spin, and with a Herculean effort kicked hard on the rudder, eased the joystick forward and realigned the DH2 with the horizon – only to discover that he was at rooftop height and about to hit the Vickers munitions factory. Too low and too slow to climb away, Mannock throttled back and brought his aeroplane down on the factory apron, only yards away from sheds stacked with high explosive. A catastrophe on a par with the explosion at nearby Silvertown later in the year was thus only just averted.[90]

Mannock insisted to McCudden that the spin was unintentional, and the only reason he had survived was because he remembered the younger pilot's advice to 'put all controls central and offer up a very short and quick prayer'.[91] A Hail Mary may have saved his life – and the lives of several hundred munitions workers – but it seemed unlikely to prevent 'temperamentally unsuitable for flying duties' being entered on Mannock's service file. McCudden exercised discretion but the other instructors made their views clear, and Swart rejected out of hand any claim that the whole incident was an accident. He made Mannock sweat for 24 hours before officially informing him that the incident was closed. Swart clearly bent the rules, and his diplomatic skills must have been severely tested in order to prevent Vickers and the Ministry of Munitions registering an official complaint to the War Office. In this respect it was in his own interest to have the incident hushed up, but there was still no pressure on him to keep Mannock. One can only surmise that his colonial background rendered him less inflexible than a British regular officer when it came to a choice between strict application of the rules and exercising discretion. Mannock was very, very lucky.

The irony is that at the time 'Micky' Mannock did not come across to either staff or students as a maverick character. In fact, exactly the opposite was the case. He scarcely drank, still attended church voluntarily, and appeared to anyone who did not know him very well as a quiet and even-tempered chap 'inclined to be almost too serious-minded'. Mannock's room-mate at Joyce Green, Meredith Thomas, would argue with him late into the night. He came to appreciate that here was a man who had an opinion on everything, but he also noted a new found tolerance: Mannock still had 'strong likes and dislikes', but he was now 'prepared to be generous to everyone in thought and deed'.[92] Major Swart was less likely to dismiss a reserved, reflective and worldly-wise apprentice pilot than a raw, immature and gung-ho candidate for early immolation. Presumably Mannock – *if* he survived those critical first few weeks – would become a stalwart of the squadron, reliable and reluctant to take unnecessary risks. Naturally, the episode with the DH2 called into question any notion that Mannock would become his squadron's pipe-smoking 'uncle' figure ever ready to advise apprehensive young fliers newly arrived in France. Anyway, Mannock was never going to conform to such a stereotyped view of the older fighter pilot, assuming of course that such characters are not confined to the most clichéd cinema representation of a front-line squadron, in either the First or Second World War.

Further solo flights in the DH2 saw Mannock complete his training as a scout pilot, and at the end of March he received orders to proceed immediately to St Omer, epicentre of RFC operations in France and Belgium. Trenchard's headquarters were located in this small but now intensely busy town, as was No. 1 Aircraft Depot, the larger of the RFC's two aircraft parks (the other was at Candas). Mannock arrived in France on 1 April 1917, exactly 12 months after he received his commission, and two years to the day after his release from internment. Ironically, St Omer reminded him of Constantinople. He loathed the dirt and the poverty, and still needed to have the puritanism knocked out of him, judging by his precious remarks regarding the quality of service entertainment on offer. After a week of being shifted from one seedy and crowded billet to another, with the occasional opportunity to take up a Bristol Scout and find out what real flying was all about, Mannock was posted to 40 Squadron.[93]

On 6 April he arrived 'cold, wet, but cheerful' at an airfield near Bruay, a grim village on the Bethune coalfield, west of Lens. 40 Squadron flew

Nieuports, agile and fast-climbing scouts highly regarded by both Ball and Bishop, but in the spring of 1917 outgunned and outmanoeuvred by the new D3 Albatros. Once again the Germans had gained the technological initiative: unable to match the Allies' rate of production, the performance of their aircraft was crucial to blunting Trenchard's strategy of a permanent offensive, in other words, fighting the air war above and beyond the enemy front line. Mannock was shown his aircraft and assigned to C Flight. His commanding officer, Major Tilney, ordered a dawn flight over the trenches so that he could become familiar with the landscape as soon as possible. After that he would commence offensive patrols, bomber escort and balloon strafing. The next morning, as he flew over the line for the very first time, immediately below him all four divisions of the Canadian Army Corps were less than 24 hours away from their murderous assault on Vimy Ridge. For the RFC the preliminary air offensive over Arras marked the beginning of 'Bloody April': in four days 75 aircraft had been lost in action and 105 pilots and observers killed. A further 56 machines had been wrecked, largely as a result of placing powerful but unstable scouts like the Bristol and the Nieuport in the hands of inexperienced pilots, most of whom would never fly again.[94] For Eddie/Pat/Jerry/Murphy/Mick – whoever he was at any particular point in time – the war had at last begun.

4
40 Squadron, 1917–18

An uncomfortable apprenticeship

When Mannock arrived at 40 Squadron the unspoken assumption was that here was another new face that would soon be gone. In the spring of 1917 the RFC was creeping towards an attrition rate of around 200 pilots a month. Aircraft losses rose proportionately.[1] The war in the air placed a premium on quality not quantity. By 1917–18 the German Air Service was numerically overwhelmed, with British factories doubling their output to over 30 000 machines in the final year of the war. Nevertheless, although the DH2 and the Sopwith Pup had allowed Ball and his comrades to regain air superiority over the British lines in the late summer of 1916, the advent of the Albatros D1 had enabled the Germans to redress the balance the following winter. Germany had again seized the technological initiative, superior performance compensating for inferior numbers.[2] Successive versions of the Albatros enabled Germany's elite pilots to exploit new tactics developed by Oswald Boelcke, and later by his protégé Manfred von Richthofen. Even in 1918 Germany still enjoyed the clear advantage of having entered the war with a well-established machine tool and precision engineering industry. German aero-engines were invariably more powerful and more reliable than their Allied counterparts.[3] Although the British clawed back air superiority in the course of 1917, the arrival of the Fokker D VII early the following year saw the pendulum swing back in favour of the *jagdstaffeln*, the elite 'hunting squads'. The Germans had a further advantage in that they were fighting a largely defensive war, preferring to attack only when enjoying a numerical advantage (the notorious 'circuses'), and when supported from below by their own anti-aircraft batteries. The RFC, particularly its scout squadrons, fought the war beyond no man's land, in accordance with Trenchard's strategy of a permanent offensive. Thus Sopwith Camels, often used to protect artillery reconnaissance aircraft, might fight up to ten miles behind enemy lines, and an SE5 patrol might penetrate perhaps twice as far. In the air, as on the ground, tactics and technology went hand in hand. Thus the superior engine capacity of German aircraft enabled them to climb fast, and cruise at high altitude, thus facilitating rapid attack followed by early escape eastwards. A large German patrol would have aircraft flying at different levels, leading Camel pilots to complain of their inability to climb high enough or fast enough to facilitate an attack from above.[4] Combat at altitude, high above enemy territory, became the prerogative of the SE5a, but only in the autumn of 1917 when Farnborough and its subsidiary factories could secure large supplies of a decent engine. For 40 Squadron the previous April this new generation of fighter aircraft remained a dim and distant prospect. Mannock would serve his apprenticeship in the increasingly outclassed and outgunned Nieuport 17, by no means a poor aircraft, but a victim of each side's relentless determination to gain that crucial technological edge.

Mannock joined his first squadron at a point in the war when the life expectancy of a subaltern from posting to death was a mere 11 days. A familiar theme in both fictional and non-fictional accounts of life in a front-line RFC squadron is the indifference of the experienced pilots to replacements fresh from England.[5] All too often the new faces were left to sort out their own messing arrangements, their arrival overlapping with the dreary ritual of signing off their predecessor: the official report, the letters of condolence, the disposal of personal effects and possessions. Trenchard's insistence on 'No empty chairs at breakfast' was intended the minimize the impact of heavy losses on squadron morale by the rapid replacement of dead friends and comrades.[6] Memories of the brave and long serving would never be erased, but there still remained those whose names could scarcely be remembered. In age the senior members of the squadron were rarely that much older than the novices, but they were the survivors, the compression of time and service leaving them old beyond their years. Mess veterans saw little point in getting to know the gauche, naive and nervous young men who, in only a few days, would likely as not be gone. If they learnt fast and were lucky then they would soon enough be accorded an individual personality – and gain the respect of their peers.

Mannock was one of the lucky few who survived those first gruelling weeks, but the achievement of staying alive had to be set against his fellow pilots' growing conviction that he became paralysed by fear from the moment his Nieuport left the ground. In other words, he was a liability to himself, and he was a liability to the rest of his flight. Earning the respect of his fellow officers was to prove an unusually long and painful process, if only because, having finally got to France, Mannock was in no rush to get killed. In his early weeks with 40 Squadron he was viewed with deep suspicion. He would soon have been sent home had he not been sensitively handled by his commanding officer. An Old Etonian and ex-yeomanry officer, Major Leonard Tilney was at first sight the antithesis of Mannock. Yet Tilney was perceptive enough to appreciate that here was an older man too intelligent to embrace the fatalistic, live-for-today mentality so aggressively projected by most young pilots as a means of coping with their private fears and nightmares.

From the moment he sat down in the chair of a squadron favourite killed that day, Mannock was a marked man. Although both his flight commander and his room-mate found him shy and unassuming, others found him just the opposite. Intrusive questions on his first day, and ill-informed statements on how to fight the Germans, were rightly deemed crass and insensitive.[7] Not only that, the brash new arrival talked with a slight Irish accent, and displayed scant respect for mess manners and conventions:

> ...he [Mannock] was different. His manner, speech and familiarity were not liked. He seemed too cocky for his experience, which was nil. His arrival at the unit was not the best way to start. New men took their time and listened to the more experienced hands; Mannock was the complete opposite. He offered ideas about everything: how the war was going, how it should be fought, the role of scout pilots, what was wrong or right with our machines. Most men in his position, by that I mean a man from his background and with his lack of fighting experience, would have shut up and earned their place in the mess.[8]

Squadron morale was rooted firmly in understatement. This implicit understanding of a common set of distinctly upper middle-class values meant that any newcomer coming from a different background had to learn the ground rules or face ostracism. In *Goodbye to All That* Robert Graves recalled how:

> Patriotism, in the trenches, was too remote a sentiment, and at once rejected as fit only for civilians, or prisoners. A new arrival who talked patriotism would soon be told to cut it out.[9]

Mannock was making a clumsy attempt to hide his nervousness, but he came across as boorish and opinionated. A readiness to discuss politics rendered him even more unpopular, particularly given his progressive agenda. Spring 1917, with MacDonald eager to meet revolutionary leaders in Petrograd and Arthur Henderson forced to leave office over Labour's presence at the Stockholm Conference, was not the best time to inform new colleagues that one was an active member of the ILP. Another member of Mannock's flight, an Irishman called de Burgh, was almost alone in wanting to take on 'the only man in the mess who would talk beneath the surface'. Bizarrely, he and Mannock would conclude their very public arguments by having a boxing match.[10] Nightly fisticuffs no doubt irritated those who felt little goodwill towards the Irish at the best of times. Far worse, however, was the fact that after only a month in France Mannock had been dismissed by some of the squadron as not only an irritating loudmouth but also a coward.

Flying an unreliable Nieuport scout in the face of fierce ground fire, let alone attack from veteran *jastas*, left the inexperienced Mannock near to nervous breakdown. After two months in France, and shortly before his first leave, he noted in his diary: 'Feeling nervy and ill during the past week. Afraid I am breaking up.'[11] Events had conspired to render him 'nervy and ill' almost every time he went up. In only his first week Mannock had seriously damaged his aircraft on landing. He had emerged unscathed, unlike an old friend from Joyce Green who was in hospital in a critical condition. His diary makes daily references to 'poor old Dunlop' who was in a coma. Mannock eventually felt the need to go and see him. This sobering experience was followed the next day by his first time over enemy lines, escorting FE8 bombers. Ira Jones, writing in the aftermath of the 'Bodyline Tour', considered this rite of passage 'even more tense than receiving the first ball in cricket from a Larwood'. But then, as befits one who played scrum-half for the RAF, Jones considered that airfighting in general was 'just like Rugger'.[12] Mannock appears to have been totally disorientated by his first acquaintance with German anti-aircraft fire – 'Archie' – and to have lost control of his machine. Once he recovered from his initial panic he linked up with one or two others who had similarly lost contact with Captain Todd, the flight commander. The ever-loyal Jones gave a charitable interpretation to this whole incident, but Dudgeon argued persuasively that it had a deeply demoralizing effect on Mannock.[13]

For the first time Mannock discovered the reality of flying in combat, the priority being to keep calm and react instinctively – neither of which he seemed capable of doing. However, the overwhelming requirement was to control a gnawing, potentially crippling fear, which if anything intensified every time he went out on patrol. Often he was sick before take off, and his diary and letters indicate that he was now drinking quite heavily. The diary also notes those pilots who arrived in France with Mannock and were already dead, in at least one case shot down in flames by the 'Red Baron' himself.[14] At this early stage Mannock felt every death keenly, and already he was scarcely able to control his fear of being burnt alive while trapped in the cockpit. The horror of dying in a 'flamer' ultimately became an obsession, but then, 'There were few flyers with any experience of air fighting who were not obsessed to some degree, though usually secretly, with the thought of being shot down in flames.'[15] For the present, however, the near traumatized new pilot had found no one whom he felt able to confide in, hence a failure to appreciate that many of his feelings were, to a greater or lesser degree, shared by all pilots on active service. On 20 April, after a particularly harrowing day, he looked back over the previous week's patrols:

> Now I can understand what a tremendous strain to the nervous system flying is. However cool a man may be there must always be more or less of a tension on the nerves under such trying conditions. When it is considered that seven out of ten forced landings are practically 'write offs', and 50 per cent are cases where the pilot is injured, one can quite understand the strain of the whole business.[16]

Letters sent home in the second half of April, during which time he was crossing the German lines at least twice a day, offer a further insight into Mannock's state of mind. He admitted to Jim Eyles that under fire for the first time he 'had the wind up', and confided to his mother that 'it's very lonely being up in the clouds all by one's self, with the anti-aircraft shells coughing and barking all around one, and big guns on the ground flashing and spitting continuously'. Writing a week later to his mother he again admitted how terrified he had been throughout these early 'aerial scraps'.[17] When in action he never felt in control, and he seemed incapable of firing straight, that is assuming that his solitary Lewis machine gun had not jammed: Mannock had been left a helpless observer in a dogfight over Douai on 1 May. Although he only just escaped with his life when attacked from behind, the claim that a jammed gun kept him out of the action provided further ammunition for the squadron's sceptics. To compound his woes, he had little confidence in an aeroplane which in his second week with the squadron literally broke up as he pulled out of a dive only 700 feet above the ground. The following day the engine of his replacement aircraft cut out three times while over the lines.[18]

Having survived a near-fatal crash, Mannock's feelings of insecurity were deepened by a growing awareness that visible signs of acute nervous tension were serving only to reinforce his fellow pilots' prejudices. However, those old hands still willing to give him the benefit of the doubt pointed out that to pull out of a dive with the lower right wing torn out of its socket, let alone get the aircraft down otherwise intact, was, after all, airmanship of high order. Mannock had been practising his gunnery over the airfield when the near-catastrophe occurred, not that this eagerness to improve his marksmanship impressed the cynics. It seemed to some that this keenness, coupled with such fine words on his first night in the mess, sat uneasily with an evident lack of aggression in the air.[19]

Mannock's fiercest critics in the squadron saw confirmation of an alleged readiness to make excuses for avoiding combat in the extreme precautions he now took when preparing his aircraft. As well as supervising the Nieuport's flight-worthiness, he informed the armourer that henceforth he would assume responsibility for sighting and loading his machine gun. Not for the last time Mannock sought advice from Major Pearce, the RFC's peripatetic armament officer. The two men established a close working relationship, their collaboration on best practice in aerial gunnery ending only with Mannock's death. By personally checking the ammunition he could identify faulty bullets, a major cause of jamming. Resetting the cowling sights enhanced accuracy at short range, ensuring a much greater chance of hitting an enemy aircraft: 'By closely observing the firing pattern the pilot could shift the gun mountings to get his sights exactly synchronised on the gun burst at a particular range.'[20] At the same time Mannock practised daily on the target range in order to improve his marksmanship, and thus overcome any deficiency he might have had in his left eye. The ground crew would drag his Nieuport over to the butts and line it up opposite the target as if in a flying position. Like other successful fighter pilots, notably Albert Ball and 'Billy' Bishop, Mannock recognized early on that maximum damage could only be inflicted on a fast-flying target through rapid, accurate fire *and* close proximity.[21] Increasing the likelihood of a first time direct hit was crucial as, not only did the Albatros outperform the Nieuport, but also it carried two cowling-mounted Spandau machine guns, and therefore had more than double the firepower of its adversary.

Through direct and painful experience Mannock was coming to appreciate the importance of reliable equipment. Age also imposed caution: to attack the enemy, yet be either too scared or too tactically inexperienced to know what to do next, signalled an early death. Mannock had the maturity and common sense to realize he had to learn quickly from initial engagements, the aim being to heighten senses and to cultivate skills no amount of advance preparation could provide. There were also longer-term essential requirements: the rapid loss of raw pilots had to be stemmed through close supervision and patient induction, and above all the need to emulate the Germans and develop effective formation tactics common to flight, squadron and even corps.[22]

The intensity of air operations is shown by the fact that Mannock flew over 40 hours on patrol in his first three weeks, with a similar rate of engagement throughout the following month. Nervous tension, damp weather and an unhealthy diet left him feeling unwell for much of this time. He told the Eyleses that, 'I always feel tired and sleepy, and I can lie down and sleep anywhere or at any time.' Constant fatigue was a given consequence of flying at high altitude. Like most scout pilots he suffered from constant headaches, a consequence of oxygen deprivation, intense concentration, stress and 'induced ocular fatigue'. Rapid changes in atmospheric pressure may also have given him earache, another common problem.[23] His emotional condition was further undermined by a succession of costly missions in which he barely escaped with his life. At the same time he became ever more despondent that, despite all his best efforts, he had been unable to shoot down a single aircraft. His misery was compounded by the knowledge that some in the squadron still believed Mannock was consciously avoiding a direct encounter with 'Jerry'. In such a repressed environment, where insouciance was always expected to triumph over neurosis, cowardice was not just 'a deadly sin', but *the* deadly sin.[24] The exultation of at last striking a blow against the enemy – by taking part in a successful balloon-strafing raid on 7 May – lasted no time at all. Flying at low level through ferocious anti-aircraft fire deep in enemy territory, firing a long burst of tracer bullets and then climbing steeply to avoid the exploding balloon, was bad enough, without being attacked by five Albatros D3s. Four of the six-man patrol crashed, and the newly arrived flight commander was shot down. Mannock was the only pilot to land his bullet-riddled machine intact. No wonder he recorded in his diary, 'I don't want to go through such an experience again'.[25]

Although he didn't know it at the time the torch had been passed – that evening Albert Ball, now flying an SE5 with the elite 56 Squadron, flew his last patrol. Ball had been living on borrowed time, and by the end was suffering from severe 'anxiety neurosis'. Twice in one week his aircraft had been severely shot up. Unlike Mannock of course he was paying back with interest: he died having downed 43 enemy aircraft. Two days before his death he wrote to his father: 'I do get tired of always living to kill and am really beginning to feel like a murderer. Shall be so pleased when I have finished.'[26] Appropriately, Ball died alone. He was by no means the last solitary warrior, and Billy Bishop, in many respects the definitive loner, survived the war. Nevertheless, Ball's final few weeks in France marked a coda to that phase of the conflict where individual genius was still a defining factor in securing air superiority. Mannock was yet to make his mark, but his legacy would be a recognizably modern form of aerial warfare based on tight organization, discipline, trust, close collaboration and, above all, teamwork.

The rumour machine worked rapidly within the RFC, and Mannock would soon have picked up on the news that the corps' talisman was dead. Ball's exploits had been one reason why he had decided to quit the Royal Engineers, but now he was having serious doubts about whether, mentally and physically, he was cut out to be a scout pilot. Mannock had survived the balloon-strafing with his nerves just about intact, but earlier that same day he had been severely shaken when a sudden attack by several 'red devils' had left him isolated and exposed. Only two days later he again found himself alone behind the German lines with three enemy aircraft on his tail, but this time his engine was spluttering and his machine gun had jammed. He dived, and dived again, dropping over 13 000 feet towards Arras before regaining power and shaking off his attackers, who now found themselves too far west. Almost immediately he found himself facing yet another Albatros, with the Lewis still disabled. Mannock quickly made his escape, landing at Bruay 'with my knees shaking and my nerves all torn to bits...all my courage seems to have gone'.[27] One look at Mannock's face must have been enough to tell Tilney that if he sent him out on a second patrol that day then he would never see him again. Shrewdly, he gave him a break, sending him off to St Omer for a replacement machine. Mannock had time to regain his composure and to think. Dudgeon saw these few brief hours of privacy as marking a watershed for his hero, and it is true that the diary entries after 9 May are less broody and more positive in tone. Having said that, when recording the events of the past few days Mannock could still speculate on whether he would break, and what would be the reaction of his friends. ('Old Paddy the "devil may care" with nerves. I feel nervous about it already.')[28]

What is clear from both Mannock's diaries and his regular letters to the Eyles family is that, for all the whisperings in the mess, by mid-May he had certainly won over the rest of his flight and would soon be established in the squadron. Lionel Blaxland wrote to Julia Mannock towards the end of his leave, reassuring her that 'Your son is very good indeed at the job, and is very much liked in the Mess.'[29] An exaggeration perhaps, but nevertheless an indication that 'Mick' was becoming more widely accepted as one of the boys. He was now regularly frequenting the hotels and *estaminets* of Brouay, Bethune and St Omer. Ironically, he had been encouraged to stop brooding, and to go out and get drunk, by 40 Squadron's chaplain, B.W. Keymer. He and Padre Keymer were to become close friends, the latter's muscular Christianity, earthy humour and unique capacity to communicate with his congregation earning him respect throughout the RFC. His popularity, and the success of the Church of St Michael ('The Flying Warrior') which he established inside a Nissen hut at Bruay, were officially recognized after the war: Keymer was the RAF's first chaplain at Cranwell. At Bruay he held meetings rather than services, so that after a few hymns and prayers pilots could relax with a cigarette while their padre delivered a brief, no-frills homily. Keymer felt he was in no position to preach, but simply be there when he was needed. He invariably joined the ground crew in anxiously seeing patrols off and enthusiastically welcoming them back, and much of his quiet counselling seems to have taken place on the tennis court. As losses mounted in the late summer of 1917 Keymer speculated upon whether he would be making a more useful contribution to the war if he left the clergy and applied to train as a pilot. It was Mannock who politely but firmly pointed out that this was an absurd suggestion, and that Keymer's supportive role was vital to squadron morale.[30]

Keymer acted as a substitute for Jim Eyles, as did 'Zulu' Lloyd, a tough South African serving with 60 Squadron when he first met Mannock in St Omer. George Lloyd's leadership qualities were recognized by his eventual promotion to major and appointment in 1918 to an SE5a training squadron at the Central Flying School. Before that, however, he spent the second half of 1917 as a flight commander in 40 Squadron. Lloyd was popular, with 'a charming manner, a sort of boyish ingenuousness that disarmed any criticism or animosity', but he also generated respect.[31] He quickly gained Mannock's confidence, acting as a confidant when the going got especially tough. In the weeks preceding Lloyd's arrival, Mannock's more obvious aggression in the air had helped to temper earlier criticism. Tilney, as a demonstration of faith and a clear signal to the rest of the squadron, more than once put him in charge of a patrol, despite the fact that Mannock had yet to claim his first victory. Now confident in both aircraft and weapon, and determined to reap the reward of daily gunnery practice, Mannock grew ever more impatient. On at least three occasions he was fairly certain that he had shot down a two-seater reconnaissance aircraft, but each time was unable to get confirmation of his 'kill'. All too well aware of latent hostility in the mess he judged it wiser not to press a claim. On 7 June, while escorting FE2b bombers over Lille, Mannock reported his first victory. He destroyed an Albatros at close range, and two days later downed another one north of von Richthofen's headquarters at Douai. Both kills were reported, but neither was confirmed, the aircraft crashing well behind enemy lines ('I felt like the victor in a cock-fight!').[32]

Any further success was interrupted by Mannock abruptly going on leave for a fortnight. He had expected his first break some time in mid-July, but events conspired to persuade Tilney that Mannock should go sooner rather than later, even at the cost of annoying anyone hoping for an early departure for Blighty. Psychologically, Mannock was by now in a bad way, displaying all the characteristics of pilots who had been living on the edge of their nerves for too long without a proper break. It was not as if there had ever been a time when he had viewed airfighting with at least a degree of equanimity. He had never conquered his fears, not least the terror generated by a very real prospect of burning to death in a conflagration of dope, oil and high octane aviation fuel. This was not the product of a troubled imagination, or one's worst nightmare, but an immediate probability awaiting any scout pilot still alive after two months at the front. Mannock would never overcome his worst fears, and by the end they would be virtually uncontrollable, but he did adopt strategies for dealing with them. He may well have served for another month in the squadron had he not suffered a particularly nasty injury when bringing his Nieuport in to land. Without goggles to protect him he felt a searing pain as a small piece of metal entered his right eye, leaving him partially blinded. Somehow he got the aircraft down, was dragged out of his cockpit in a state of shock, and promptly fainted in the mess as soon as comrades tried to extract the offending object. Taken to hospital, Mannock was given cocaine before a further attempt was made to clean up his eye. The pain of the throbbing, bloodshot eye kept him awake all night, necessitating a further visit to St Omer. 'More teeth-gritting and profanity' saw a second steel fragment removed, and yet still the pain refused to abate. A third operation was necessary before the area around Mannock's heavily inflamed eye was finally clear. Denied vital peripheral vision, and with one or both eyes aching, Mannock's search for the elusive confirmed kill was temporarily at an end. By this time, however, his diary makes clear that he was desperate for a break. He was put on light duties, or let off flying altogether, and in the end he was sent home. On 17 June he departed Bruay for Boulogne.[33]

Note by the way that in the week following his accident Mannock flew on several occasions, visiting other squadrons and collecting a spare machine from St Omer. He may not have flown until the inflammation had gone down. But if he did fly when he still had a dressing on his right eye, the vision in his left must have been pretty good otherwise how would he have been able to see? Interestingly, William MacLanachan – 'McScotch' – insisted that in late August a similar incident occurred when Mannock landed far too fast and crashed into a haystack. Because the landing was in a field near the front line MacLanachan was the only witness. As noted in Chapter 2, 'McScotch' claimed in *Fighter Pilot* that Mannock smashed the right side of his face into the windscreen when his aircraft first made contact with the ground – an engine seizure had forced him to glide down from over 10 000 feet. Unable to see clearly Mannock failed to avoid the haystack, and yet in such a confined space and with no engine he would have had difficulty halting the aeroplane anyway. He recorded the incident in his diary, but made no reference to any problem with his eyesight. A later passage in *Fighter Pilot* formed the basis for the myth of the 'ace with one eye', as all subsequent accounts are based on the secret supposedly shared with MacLanachan by his flight commander, namely that he always arrived early for eyesight tests and endeavoured to learn all the letters on the Snellen chart.[34] Chapter 2 questioned the credibility of this story, and Mannock's earlier accident can be seen as further evidence that, whatever problem he *might* have had with his left eye, it did not amount to total blindness.

In England the dutiful son first visited his mother. Julia Mannock had moved to Birmingham from Belfast. Perhaps she left Ulster because she couldn't endure seeing her eldest daughter struggle to keep her marriage intact.[35] Just as Jess was having a hard time, so too was Norah in Kent, and neither daughter could have felt happy about having to support a mother who was now drinking heavily. Patrick was serving with the Tank Corps in France, and not being an officer he found it much harder to secure home leave. Having married in 1915 he could not be expected to support his wife and his mother on a sergeant's wage. No doubt Julia put heavy pressure on Edward to make sure she was well provided for, but the thought of any money sent being spent on drink must have been a strong disincentive. Anyway, Mannock felt increasingly distant from his mother, his letters to the Eyleses repeatedly making it clear that home was 183 Mill Road. Not surprisingly, therefore, he soon said farewell to Julia, and caught the train to Wellingborough.

Mannock gave the Eyleses a full and honest account of his experiences over the past three months. He spent hours explaining to Jim Eyles the nature of airfighting and the need for new ideas. If his friend ever lost the thread of the argument then Mannock would get quite irate. The Eyles family had never seen him so totally immersed in any one subject, nor so enthusiastic. Pat could be found standing on the table, playing out untried methods of attack, with Jim putting on a brave face – or keeping a straight face – as he pottered around the kitchen waiting to be shot down. Mannock acknowledged that age was not on his side, but felt sure he could compensate for 'the weakness of the flesh' by spending long periods of time in concentrated thought, working out fresh tactics in his head – hard thinking would reap its reward once he came to put theory into practice ('You watch me bowl them over when I return!').[36]

Leaving Wellingborough Mannock caught the train to London, where he spent a few days at the RFC Club in Bruton Street, just off Berkeley Square. Ira Jones insisted that he was the Club's most popular member, which was carrying hero-worship perhaps a little too far, but he was certainly the life and soul of the party. While he was there a 'rag' got rather out of hand, and the following morning a collection had to be taken to pay for running repairs and wrecked furniture. For the remaining 12 months of his life Mannock could when necessary pose as an unashamed extrovert, his immersion in mad escapades and wild sprees a convenient means of releasing tension and camouflaging inner agonies. Mess rags not only confirmed Mannock's reputation from the autumn of 1917 as a thoroughly 'stout fellow', but in his eyes were well worth initiating if they went some way to maintaining mess morale in the face of imminent death. In this respect he had much in common with 'Boom' Trenchard, even though Mannock disliked his general intensely from the first time they met. However, he must have made an impression, as from the summer of 1917 Trenchard began to take an interest in perhaps the least conventional of his front-line pilots.[37]

Before returning to France Mannock stayed a few days in Barnes. He stopped with his cousin, Patrick, who was editing *Whitecraft*, the company magazine of Whitehead Aircraft, a small family business based in nearby Richmond. Patrick's politics were similarly left-leaning, witness his many years on the staff of the *Daily Herald*, and Whitehead's singularly inappropriate choice of the pro-Labour *Reynold's News* in which to advertise for fresh investors. Edward was by no means dismissive of his cousin's attempts to write about aeroplanes, and a year later offered to contribute 'some "real live stuff"', provided that he was suitably reimbursed. Patrick's father was of course the billiard-playing J.P. Mannock, and the two of them insisted on taking the poor relation for a night out on the town. Cousin Edward belatedly became acquainted with the world his father had deliberately turned his back on, and he was singularly unimpressed. All his prejudices appeared to be confirmed, and he made disparaging remarks to Patrick about aristocrats not deserving deference simply because of 'a handle to their names'.[38] Mannock's antipathy towards the aristocracy was virulent. Every setback in the war effort was seen as further evidence of an indolent, self-indulgent elite needlessly sacrificing lives through an inability to focus upon anything other than trivia. Although in the last year of his life Mannock became a familiar figure in Bruton Street, he frequently railed against clubland and London Society, having a particular prejudice against titled hostesses.[39] Almost the first question he asked the scholarly William MacLanachan when he joined 40 Squadron was, 'Are you a snob', explaining:

> '...snobbery is a nasty word, I only apply it to social and money snobbery – the empty social type particularly. There are so many damned social snobs about.'[40]

MacLanachan was of course recalling Mannock's remarks 19 years later, but the subsequent interrogation was clearly intended to 'estimate my character and my spirit, not my social position':

> 'That's it, old boy. It isn't the school or the university, nor who your father is that matters, it's what you've got in your head and your guts.'[41]

Other pilots recalled similar conversations, drawing attention to Mannock's keen sense of a lack of a formal education.

Despite the uniformed drones from Oxbridge who, from his arrival at Fenny Stratford until the day he died, regularly drove him to despair, Mannock had an idealized view of university. This healthy respect for scholarship was fostered by the Eyleses, and reinforced by acquaintance with pilots like the Spinoza-reading MacLanachan and the eccentric classicist Rhys Davids.[42] Mannock's enthusiasm for education reflected a firm belief that socialism was, if nothing else, about maximizing opportunity for all: the war, for all the efforts of the 'snobs' to preserve the status quo, offered the chance for talent and expertise to triumph over birth and position. The RFC, with its pilots drawn from the 'university of life' and from all four corners of the Empire, was a microcosm of Mannock's ideal of a meritocratic war effort – at the end of the day, results were the only judge. Conversely, those institutions which were rooted in privilege and tradition had all too often been found wanting:

> The supposed 'pride' in family, school, university, and nationality of hundreds had been exposed as worthless by fellows who used their social positions to obtain comfortable service jobs at home or safe billets behind the lines overseas.[43]

This was the familiar exaggerated moan about staff officers, albeit given an ideological spin. Presumably a Guards subaltern leading his platoon over the top did not incur Mannock's wrath, however lengthy his family tree and expensive his education. The RFC was full of similar types, but the war in the air was a levelling experience.

'New Liberalism' would have meant little to Mannock, but his faith in meritocracy and positive freedom was worthy of T. H. Green. Unashamed of his proletarian origins, he grasped at a genuine opportunity for upward social mobility, hence the trip to Turkey, with its alluring appeal of both eastern romance and a decent job. Yet of course Mannock was anything but a social climber. In Constantinople he had always remained somewhat detached from the British community, and from that first night out with his uncle and cousin he regularly snubbed anyone eager to lionize a glamorous fighter pilot.[44] Cecil Lewis later recalled how even the soberest officer could be affected by the desire of so many civilians to romanticize life in the Royal Flying Corps:

> Flying was still something of a miracle. We who practised it were thought very brave, very daring, very gallant: we belonged to a world apart. In certain respects it was true, and though I do not think we traded on this adulation, we could not but be conscious of it.[45]

Mannock could scarcely avoid London's fascination with the knights of the air, particularly once responsibility for the safety of the capital lay with some of the RFC's most distinguished fliers.[46] Yet, he resolutely refused to let his head be turned. Rather than graciously accept flattery, he was if anything extremely rude.

Nevertheless, MacLanachan was convinced that Mannock returned from leave less eager to wear his politics on his sleeve, preferring to propagandize by example: 'We've got to get on with the work and leave the retribution until afterwards...'[47] Whether he did or didn't say words to this effect, the one-time secretary of the Wellingborough ILP was from now on even more focused on the job in hand. He was keen to put his new ideas into practice, and he was eager to impress 'Zulu' Lloyd. Lloyd's relationship to Mannock was similar to the latter's friendship with MacLanachan. When 'McScotch' had arrived at Bruay in late May, Mannock had immediately taken him under his wing, despite the fact that he himself had only been on operational duty for a few weeks. It was Mannock who later gave the Scot his nickname, to distinguish him from another close friend in 40 Squadron, George McElroy or 'McIrish'. Mannock protected MacLanachan when he was attracting almost as much animosity from the rest of the squadron as his newly acquired mentor. Ironically, while Mannock was on leave the younger man had established his airfighting credentials, silencing even the harshest critics in the mess. By this time criticism of Mannock himself mainly sprang from two pilots: one of the four South Africans in the squadron, and an aggressive, highly successful and extremely opinionated Canadian called Steve Godfrey. Hall, the South African and Godfrey even went so far as advising 'McScotch' to distance himself from Mannock, otherwise he too would start to 'stay out' whenever the going got tough.[48]

Mess morale was, if by no means poor, not as high as it needed to be at yet another crucial juncture in the war. To the north, the Messines ridge had already been taken, with the RFC playing a major support role, and the main assault at Ypres was due any day (it actually commenced on 31 July, ending inconclusively in the mud of Passchendaele four months later). Godfrey's ill feelings towards Mannock may well have been compounded by the speed with which he was awarded the Military Cross. Mannock heard the news on 19 July, not that long after the far more experienced Godfrey had learnt of his award. Was there a feeling among the more experienced pilots that their pride in winning the MC had somehow been diminished by the apparent ease with which this relatively new arrival had secured his first medal? Even the citation in the *London Gazette* tacitly acknowledged that Mannock had not secured any confirmed kills at the time his 'conspicuous gallantry, and devotion to duty' had earned him Tilney's recommendation and Trenchard's approval. Special mention was made of his part in the downing of three balloons, but the same high level of courage and determination had been shown by everyone who had flown on these missions.[49] Thus, was the CO's decision to encourage Mannock, and to recognize his potential, perhaps premature? In terms of the squadron as a whole, was the gesture counter-productive, despite the fact that Mannock was an acting flight commander and had shot down two more reconnaissance aircraft by the time confirmation of his MC arrived? He may have registered two more kills by the end of July, but by his own admission he had in poor visibility 'fired a drum of ammunition into one of our own machines', and the previous day had written off his own aircraft, albeit through no fault of his own.[50]

Any adverse reaction to news of Mannock's MC may have been compounded by the way he responded to his first visit to the trenches, and his first close inspection of a 'kill'. On 13 July he insisted on examining the remains of the DFW two-seater he had shot down south of Avion, and was horrified by what he saw:

> I shot the pilot in three places and wounded the observer in the side. The machine was smashed to pieces and a little black-and-tan dog which was with the observer (a captain) was also killed. The observer escaped death, although the machine fell about 9,000 feet. The pilot was horribly mutilated.[51]

Mannock's diary entry revealed far more about his feelings than had his message to Jim Eyles. The tone of the letter was resolutely jaunty, despite mention of the pilot's broken body. Enclosed was a piece of fabric from the downed aircraft, with a promise of further souvenirs to follow.[52] In his diary, however, Mannock confessed to being nauseated by what he had discovered in the trenches, and shocked by the sight and the smell of so many rotting bodies, not least the mangled remains of the pilot: 'I felt exactly like a murderer.' He made the same point to MacLanachan, adding that his fury at the waste of life was compounded by the fact that the dead pilot was an NCO and the observer an officer. If the latter had died instead then he would not have felt so upset – the class war had not been entirely forgotten. Despite efforts to disguise his feelings, 'McScotch' was by no means alone in recognizing Mannock's inability to take his visit to the crash site in his stride. If anything he compounded a deep sense of unease by insisting on exploring other trenches captured by the Canadians over the previous four months.[53]

Those in the mess who were still sure he was 'soft', even 'yellow', pointed to Mannock's mood swings as further evidence of his emotional instability. The loss of the immensely popular Bond, and the news that not only had Mannock been awarded the MC, but that henceforth he would command A flight with the rank of temporary captain, brought matters to a head.[54] Even MacLanachan 'wondered if promotion would prove too much for his mental equilibrium', and one or two stormy encounters over the next few days left him still unsure as to whether Mannock was fit for the job. In *Fighter Pilot* MacLanachan also hinted that a further cause of ill-will was irritation with the over-protective and proprietorial attitude Mannock had adopted towards the daughter of the owner of the *estaminet* in Bruay. Odette was a beautiful blonde woman at least ten years younger than her *aviateur* admirer, with whom she appears to have conducted a largely platonic affair. Her father had no objection, and indeed encouraged their *liaisons*, if only because Mannock would order a plentiful supply of champagne laced with brandy whenever he set out to impress Odette with his pidgin French. This apparently non-sexual relationship seemingly irritated some of 40 Squadron's less gallant and more forward 'wild colonial boys'.[55]

Major Tilney clearly needed to stamp out the factionalism and resentment currently undermining squadron morale. Yet the officers he needed to exert his authority over were precisely those who were most critical of him – for promoting Mick Mannock over their heads. Tilney therefore turned to 'Zulu' Lloyd to bang heads together. Lloyd was held in high regard well beyond 40 Squadron, and soon after his arrival he had taken it upon himself to find out why Mannock had such a remarkable capacity to annoy:

> He was not actually called 'yellow', but many secret murmurings of an unsavoury nature reached my ears. I was told that he had been in the Squadron two months, and that he had only shot down one single Hun out of control, and that he showed signs of being over-careful during engagements. He was further accused of being continually in the air practising aerial gunnery as pretence of keenness. In other words, the innuendo was that he was suffering from 'cold feet'.[56]

Mannock was remarkably frank when Lloyd asked him outright whether or not he was scared. He acknowledged that prior to going on leave he had been terrified every time he went out on patrol, but insisted that he had come to terms with his fear of death. Such a claim was open to question, particularly as the nightmare of a 'flamer' grew ever more real and intense, but Lloyd was convinced that Mannock was no coward. Not only that, but here was someone who genuinely believed he could influence the conduct of the war in the air. Mannock was older than Lloyd, but his experience was minimal compared to a man who had flown with Ball, survived the Somme and 'Bloody April', and been one of the first to test the SE5 in combat. Nevertheless, Lloyd took Mannock at face value when he claimed to have deliberately avoided taking risks in order to study the 'science' of airfighting:

> I want to master the tactics first. The present bald-headed tactics should be replaced by well-thought-out ones. I cannot see any reason why we should not sweep the Hun right out of the sky.[57]

Persuaded that Tilney was right, and that Mannock had the brains and now the temperament, to lead from the front, Lloyd set out to persuade the rest of the mess. The other South Africans and the Canadians would only be convinced once the golden boy started shooting down aircraft regularly, but most of the others had already been won over. It is arguable whether, even after Mannock was well into his remarkable run of securing 16 confirmed or unconfirmed kills in August–September 1917, everyone in the squadron had been won over. In at least one case (almost certainly Hall, lusting after Odette) the level of animosity was such that it took a blazing row to clear the air.[58]

Flight commander

'A' flight comprised 'McScotch', J. H. Tudhope – a fast-learning new arrival who quickly gained Mannock's respect and affection – and two young Canadians called Kennedy and Harrison. Although Mannock, after a hesitant start, eventually moulded all of them into a cohesive and effective fighting unit, attention focused upon a core of the flight commander, MacLanachan, and 'Tud', whose credentials as a 'particularly brave scrapper' were quickly established. Kennedy, a student from Toronto, was shot down within a month, and Harrison, being very like Mannock back in April–May, was treated sensitively by a sympathetic Mick. The three pilots at the heart of the flight became close friends, for much of the time operating individually or jointly out of an Advanced Landing Ground (ALG) less than two miles from the reserve trenches. Bruay was 11 miles from the front, but 'Mazingarbe', a large clover field with just a camouflaged tent and a hut, facilitated a rapid response to sightings of enemy reconnaissance aircraft surveying the Allied lines. Access to the ALG encouraged Lloyd, Godfrey and Mannock to spend their spare time 'hawking' for a hapless DFW or an unsuspecting Albatros cruising too low for its own good. Similarly intent on boosting his personal score, MacLanachan also began flying solo flights, in addition to normal patrols and assigned missions. The more experienced pilots preferred to operate out of Mazingarbe as they could rely on their own initiative, and were not subject to regular debriefings. An added bonus was close proximity to Bethune, where Mick and his men would relax in a teashop near the Officer's Club, eating pastries and flirting with its attractive hostess, nicknamed the 'Queen of Sheba'.[59]

At this early stage Mannock's determination to establish his airfighting credentials, and earn unqualified respect among his peers, undermined his so far paltry attempt to turn A flight into a genuine team, with common objectives and each pilot responsible for the others' wellbeing. He may have evolved new ideas for shooting down Germans, but he had yet to establish clearly in his mind how he could weld the men under him into an efficient fighting force. MacLanachan put this failure down to inexperience, a still dominant desire for individual success and, above all, emotional instability: still unable to accept that he had become a trained killer, Mannock chose 'to fight out his terrific inward battle alone', even if it meant alienating his closest comrades. Moody and raw, the flight commander chose to grieve alone, and too often he chose to fight alone.[60] Was the answer a lot simpler than this, namely that Mannock still found difficulty in articulating his ideas, and thus could not impose his will upon the whole flight? In consequence, he turned inwards, endeavouring to solve his problems alone.

If any particular incident forced a change of attitude then it was A flight's mid-August attack on Dorignies airfield. Before that, however, MacLanachan had more than once complained of Mannock risking the lives of the men under him. The flight was being led into dangerous and exposed situations, simply to satisfy its commander's appetite for kills and glory. MacLanachan applauded individual feats, most notably his friend's success in forcing down the highly experienced and decorated scout pilot, Joachim von Bertrab, and in hunting the deadly 'purple man' (almost certainly the great German air ace, Werner Voss). He shared Mannock's delight in receiving accolades from both generals and ground troops for downing von Bertrab. Not only that, he listened to Mannock when the older man urged him not to seek revenge for the death of a highly popular young pilot by loading a forbidden mixture of 'dirty' incendiary and armour-piercing ammunition. Nevertheless, he was convinced that, for all his apparent maturity of judgement, Mannock was still capable of acting in a reckless and irresponsible fashion. Although he agreed to go, MacLanachan questioned the raid on Dorignies on 13 August, pointing out the absence of any evidence that von Richthofen regularly used the airfield. Mannock was enjoying a particularly good dinner in Amiens at the time he proposed the 'stunt', but when stone cold sober the following morning he still took off. He, Tudhope and

Kennedy never reached their objective, nearly writing off their aircraft and barely escaping with their lives. A gung-ho flight commander had attacked nine Albatros, naively assuming that offensive action would in itself intimidate the enemy pilots. In fact, the Germans were highly experienced and eager for battle, and Mannock's patrol was extremely lucky to shake them off and limp back to Bruay. According to MacLanachan, a chastened Mannock was deeply shaken by the whole experience. There were two clear lessons to be learnt: firstly, that the courage and skill of the German pilots should never be underestimated, and secondly, that as a leader of men Mannock had to place the interests of the flight before individual ambition. The first responsibility of a flight commander was to protect his pilots, not boost his personal tally of kills. Although he remained as aggressive as ever, and assumed 'McScotch' would keep an eye on the others, Mannock changed his tactics to ensure the flight operated as a genuine unit. Caution now prevailed, the twin objectives being to protect the most inexperienced members of the patrol, and to ensure that the advantage always lay with the attacker. A flight now regularly practised formation attacks, with Mannock and MacLanachan guarding the two Canadians, and Tudhope bringing up the rear. At the end of the day a pilot's individual skill still remained paramount, but collective action could eliminate unnecessary risk.[61]

Although still plagued by engine seizures and crash landings, Mannock was now enjoying regular success over the German lines, the hours of gunnery practice finally bearing fruit. He noted in his diary that, 'My nerves seem better lately', and even when he did write off his aircraft he 'soon got over it'.[62] Mannock was by no means fearless, but he now had the technical ability, the confidence and the tactical nous to make him a match for all but the most experienced *jasta* pilot. Although the majority of his victims were two-seaters, he demonstrated that, in the right hands and when pushed to the absolute limit, the Nieuport was still capable of taking on the latest mark of the Albatros. One reason why Mannock experienced frequent engine failure was because he was taking the Nieuport up to a vantage point of around 20 000 feet, beyond the ceiling of even the modified versions which began to arrive at Bruay in late summer.[63] Mannock's personal success reflected upon the other members of A flight, with the exception of Harrison who was no admirer of his leader and who anyway had yet to score a victory.[64] Nor did the flight operate in isolation, with Mannock and Zulu Lloyd liaising over Mazingarbe's valuable role as a forward attack base.

However, tensions still existed across the squadron as a whole, and even within A flight, as confirmed by a number of incidents in late

August and early September. This was a period when 40 Squadron's primary role was to support the Canadians' assault on the notorious Hill 70 at Loos, and a parallel attempt to recapture Lens. Pilots were in the air for long periods of time, scanning the sky for enemy aircraft, and checking below for a warning signal laid out on the field at Mazingarbe. Although some officers had a remarkable capacity for switching off when back at the ALG – Mannock was renowned for his ability to daydream for long periods, or alternatively fall asleep almost instantly – fatigue levels rose and tempers became frayed. One incident was provoked by Lloyd and Mannock's successful attack on five two-seater trainers. When the patrol landed the two men could only placate an outraged MacLanachan by convincing him that they were genuinely unaware the 'murdered' aircrews were unarmed. All agreed that the incident should be hushed up, and such ungentlemanly 'Hunnish' conduct, albeit unintentional, not recorded in each pilot's combat report.[65] A few days later Mannock and 'McScotch' almost came to blows when the latter accused him of abandoning the flight's mission to escort bombers over Lens as soon as he saw the opportunity to join a dogfight. They finally made their peace, with Mannock acknowledging that more detailed advance planning was required before any patrol took off. Henceforth every contingency would be planned for, with every member of the flight aware of his particular role across a variety of different scenarios.[66]

Kennedy's death on 22 August provoked a full-scale row in the orderly room once pilots sat down to write their combat reports. Godfrey and Hall accused Mannock of recklessly attacking five or more Albatros above Douai, and not waiting until their flight could ensure superior numbers. This was a justified grievance, but implicit in their criticism was that he had caused the unnecessary death of one of his men. Mannock was infuriated by the charge, particularly as he had relished the few short weeks spent in Kennedy's company. Kennedy was youthful, urbane and – as a graduate of Toronto University – a fluent French-speaker whose flirtatious exchanges with an admiring 'Queen of Sheba' no doubt reinforced Mannock's conviction that only a more open, less hidebound society like Canada could produce enlightened sophisticates who were anything but 'snobs'. Already resentful of any suggestion that he had caused Kennedy's death, he argued that he had no alternative but to attack, otherwise MacLanachan would have suffered the same fate. The only reason so many aircraft were in the area was because 'McScotch' was on a solo reconnaissance mission, and Mannock had insisted the squadron provide protection, albeit from a distance. Once the solitary Nieuport had flown over Dorignies, and confirmed an intelligence report that the airfield had been abandoned, it had quickly run into the five enemy fighters. Mannock insisted to Godfrey and Hall that he had to risk the lives of his flight, or else his number two would almost certainly have been shot down. He refrained from pointing out that, notwithstanding Kennedy's death, the squadron had secured several victories, he himself claiming his eighth Albatros.[67] MacLanachan's abiding memory of the whole incident was the moment he passed over Mannock's aircraft at the height of the battle: 'I shall never forget the scared expression on his face as he instinctively cowered.'[68] He could control his emotions, but like any other human being under acute stress, he could scarcely suppress them.

No doubt Godfrey and Hall fumed as Mannock found himself the centre of attention when three American war correspondents visited Bruay soon after Kennedy's death. The ground crew of A Flight had hung the top wings of von Bertrab's Albatros in the hangar, painting on details of how the machine had been forced down. The trophy caught the journalists' attention, and Tilney insisted Mannock provide them with a story. Copy was duly filed and syndicated, and yet another intrepid British aviator found himself a minor celebrity in the American press.[69] At home of course Mannock was yet to attract the attention of the national newspapers, RFC senior staff still preferring to play down coverage of individual exploits.

Early September marked the zenith of Mannock's career as a solo 'killer', the rest of his tour of duty with 40 Squadron seeing far fewer victories and an even greater emphasis upon tactical collaboration. On 4 September in the course of a single day he destroyed three two-seaters and forced down an Albatros. Any sense of triumphalism was tempered in the morning by the chilling sight of a dead observer hanging out of the cockpit as his pilot struggled to regain control of a dying machine, and in the afternoon by the screams, stench and smoke of his 'first flamerino'. Exultant at the manner in which he had stalked his prey, Mannock then stared aghast as flames engulfed Lt Fritz Frech's fatally wounded DFW. 'It was a horrible sight and made me feel sick.' A fruitless search for the burnt-out fuselage in the scarred fields around Petit Vimy proved anything but cathartic.[70] Mannock had finally inflicted upon the enemy a mode of death that haunted him night and day. He could square his conscience, but he couldn't ignore the cruel irony or the horror. In fact Mannock never would shake off this near paralysing fear of being burnt alive, hence his claim that to avoid just such a fate he always flew with a loaded revolver. Unlike McCudden he could not convince himself that he was destroying aircraft not men, and a close examination of charred victims served only to reinforce a growing obsession with death in a 'flamer'. The macabre, manic behaviour and the elaborate black jokes he resorted to in the final months of his life provided scant psychological relief, serving only to worsen his situation.[71]

Mannock and his flight continued to log long hours on patrol, but the number of victories dropped significantly. The same was true for Lloyd, Keen, Godfrey and the other senior pilots, including MacLanachan, who by this time was well on the way to securing 20 kills in six months. According to 'McScotch' the Germans were increasingly reluctant to engage in large dogfights and so kept out of range. Unfortunately, the underpowered Nieuports were unable to catch up with the enemy before they retreated eastwards. Mannock enjoyed a short break in late October, shortly after receiving a bar to his MC, but back in Bruay he complained to Jim Eyles: 'business is very slow since I returned from leave, and nothing doing in the way of increasing the total.'[72] This left A flight with a dilemma: the general consensus was that the Germans were only interested in 'scrapping' with pilots flying alone, and yet Mannock was trying to pioneer new methods of attack based on close formation flying. He wanted to foster teamwork, but other than via ground and balloon-strafing he was being given precious little opportunity to put his ideas into practice. Mannock's insistence on a collective, collaborative effort extended beyond his flight and their ground crew, or even the squadron. He spent much of his spare time going up and down the line getting to know those units with whom the RFC operated most closely, notably the artillerymen, the 'Archie' gunners, the balloon observers and the assault troops in the vanguard of the Canadian Army Corps. CAC officers were invited to eat and get drunk in the Bruay mess, now refurbished as a mock shooting lodge. Mannock would sober the guests up by a plaintive rendering of Schubert's 'Caprice'.[73]

However moving guests found Mick's violin solos, the real star of the mess was the newly arrived balladeer, George McElroy. 'McIrish' had taken Kennedy's place in A flight, and for the first few weeks Mannock tried to keep him out of the worst fighting. The novice quickly demonstrated that he was a natural. MacLanachan recalled McElroy rarely taking unnecessary risks, even though 'his attitude towards the war was that of a terrier that has been let loose in a rat-infested barn'. However, another pilot, William Douglas, insisted that the Irishman was in fact 'very headstrong', and would soon have been killed had it not been for Mannock's strict discipline and patient tuition.[74]

'McScotch' implied that McElroy was young and newly arrived in France.[75] In fact he was 24 and had served with the BEF as a motor dispatch rider in the Royal Engineers from the autumn of 1914 to May 1915 when he secured a permanent commission in the Royal Irish Regiment. For the next year McElroy divided his time between overseas and home postings before kicking his heels as a 'gentleman cadet' at the RMA Woolwich from June 1916 to February 1917. Feeling the war was passing him by, he sought a transfer from the Royal Garrison Artillery to the RFC. Although McElroy joined his first squadron already well acquainted with the Western Front, his military career paralleled Mannock's in at least two respects – neither man cared that much about regimental/corps loyalty and tradition, but each had from the outset felt comfortable with the machinery of war. Although both men had held clerical posts in civilian life, there the comparison ended. McElroy, as befitted the son of a Donnybrook schoolteacher, was well-educated and a civil servant.[76] One can only speculate as to his politics, but *if* he supported Home Rule then his upbringing and career would suggest a more tempered embrace of nationalism than Mannock, whose enthusiasm for an 'Irish Ireland' remained firmly rooted in emotion rather than direct experience. Pat/Paddy Mannock encouraged the assumption that he was Irish, and genuinely considered himself to be so, but he never had the opportunity to explore his Ballincollig roots, or visit his sister and mother during their brief sojourn in wartime west Belfast. For all the soft lilt in his voice, he was an 'Irishman' by sentiment rather than reality. George McElroy was Dublin born and bred, hence his claim to be Ireland's most successful fighter pilot of the First World War: he was awarded the MC and bar, and the DFC twice, and his score stood at 46 victories by the time he was shot down on 31 July 1918. McElroy died only five days after Mannock, with whom he remained in close contact after the latter left 40 Squadron in January 1918. Despite warning each other not to fly too low and take unnecessary risks, both men in their final days were increasingly reckless. In his second tour of duty with 40 Squadron the now Major McElroy destroyed 16 aircraft and several balloons in the frenetic final five weeks of his life.[77]

Like his flight commander, McElroy had been a slow starter. An absence of easy targets, poor weather, problems with the new SE5a and Mannock's insistence that he take no risks left him without a victory after three months at the front. When he did at last shoot down his first Albatros Mannock good-humouredly bawled him out for not adhering to A flight's contingency plan for attacking a superior force. Whether it was his mentor's departure, or, as Gwilym Lewis observed, that he had 'suddenly got his eye in and gone right ahead', but by the time McElroy was posted south to 24 Squadron in mid-February 1918 he had destroyed a dozen 'perfectly good Huns'.[78] Sent home three months later to instruct, he followed Mannock's example earlier in the year and kicked up such a fuss that within weeks he was back as a flight commander at Bruay.

Unlike MacLanachan, who looked to veteran airfighters like Zulu Lloyd for instruction, and to Mannock for inspiration, McElroy 'was the first of Mick's pupils in the true sense'.[79] Tudhope was very much his own man, and Harrison was forever suspicious of his flight commander's intentions. Mannock had previously focused upon systematic training of the whole flight, but now concentrated upon the individual. By establishing a close working and personal relationship with McElroy he could instil in him the ethos of teamwork and shared best practice. Nor would this be a one-off experiment, as Mannock was intent on improving the combat readiness of all pilots posted straight from England to a front-line scout squadron. A wider application of these ideas would prevent a repetition of his own early experience, and might even reduce the alarming attrition rate for new arrivals.[80]

Mannock's final two months with 40 Squadron were marked by congeniality on the ground and tension in the sky. Godfrey and Hall had been posted to squadrons protecting London from the Gotha bombers, while Gwilym Lewis arrived as a replacement flight commander. Lewis, whose youth belied his experience, got on well with Mannock from the moment he arrived at Bruay. Although there was a difference of ten years in their ages they shared the same systematic approach to formation flying: Lewis later boasted that in eight months with 40 Squadron he never lost a single novice pilot in combat. He quickly became an ardent admirer of his new squadron's 'expert Hun-strafer', and clearly regarded his emotional turmoil as a thing of the past.[81]

However relaxed Mannock may have appeared to Gwilym Lewis, Tilney and MacLanachan spotted tell-tale signs that all was not well. A growing preoccupation with premonitions – not least a firm conviction that, unless he shot himself first, one day he would die screaming in an aerial inferno – signalled the need for a break. It was time for Mannock to go home, even if only temporarily, but the squadron commander was unable to secure an early posting to Home Establishment. Resistance came from Mannock himself and from senior staff. Tilney's superiors were acutely aware of the need to retain experienced pilots once Haig launched his surprise offensive towards Cambrai on 20 November 1917. RFC scouts were deployed in low-level bombing and ground-strafing roles, with over 300 aircraft and around 380 tanks focused on punching a hole in the Hindenburg Line. 40 Squadron was not part of III Brigade RFC's concentrated force, but nevertheless fulfilled a complementary role north of the battle. The Germans needed to counter repeated strafing of their troops as they moved across open country to stem the British advance. As a result von Richthofen's *jagdstaffeln* was rapidly moved south, leaving the skies east of Vimy Ridge empty of 'red devils'.[82] At the same time the weather, already poor, deteriorated sharply. The rain, which further north had already created the muddy swamps of the Ypres salient, prevented either side from maintaining regular patrols. The absence of a routine was deemed to erode each pilot's alertness and combat readiness, rendering everyone in the squadron just that little bit more jumpy.[83]

The true state of Mannock's nerves became clear to his colleagues when he overreacted to the disappointing performance of the Nieuport 17's successor, the SE5a. One reason for Lewis's posting to 40 Squadron was that he had spent six months at the CFS training pilots to fly the next generation scout. He knew from experience the SE5's limitations, as well as its potential, unlike Mannock who had been forced to rely on second-hand reports. The pilots of 40 Squadron knew that there had been problems – why else had Ball insisted on alternating between the SE5 and his trusty old Nieuport when posted to 56 Squadron?[84] Nevertheless, there were high expectations of the modified SE5a, with Mannock no exception. Experimenting with extended high altitude flying he had pushed the Nieuport well beyond its ceiling, and was eager for an aircraft sturdy enough to climb above 20 000 feet and then stay there.[85] The other great attraction of course was that the Lewis was now complemented by a cowling-mounted Vickers machine gun capable of firing up to 500 rounds. What pilots and their armourers hadn't appreciated was that, because it was firing through the propeller, the Vickers was appreciably slower than the Lewis. However, the Constantinesco gearing was often badly fitted, so that the engine speed and the Vickers' rate of fire were not synchronized – bullets were failing to pass freely through the propeller's arc. The SE5a's Vickers had a tendency to jam, but now so too did the Lewis. For anyone as fastidious and attentive to detail as Mannock, even a hint of unreliability in his armament was infuriating. He wrote to the Eyles expressing his annoyance at the way easy targets were escaping because both his guns had jammed. Damaged propellers and non-firing machine guns were bad enough, but the principal reason why 40 Squadron failed to destroy a single enemy aircraft during its first three weeks with the SE5a was the unreliability of the 200 h.p. Hispano

Suiza engine. A number of factors contributed to a level of engine seizure way above the accepted norm: flawed design, particularly with regard to lubrication and transmission; poor machine work by subcontractors not equipped for precision engineering; an acute shortage of spares; and appalling quality control. None of these problems were properly addressed until production switched to a modified version of the original, the Wolseley Viper. Mannock's aircraft in 74 Squadron were fitted with this marginally more reliable engine. Nevertheless, for the remainder of his flying career he was primarily dependent upon his own technical skill, and the expertise of ground crew on a steep learning curve, to ensure that the teething troubles of Christmas 1917 did not persist.[86]

Problems with the Vickers eased after a gunnery officer was sent from St Omer to advise the fitters and armourers on modifications to the synchronizing gear. At the same time he ordered all guns to be realigned (to ensure the trajectories met), and to be regreased (the existing lubrication was freezing at operational height in winter temperatures). However, when this armaments expert first arrived Mannock berated him, menacing the hapless envoy from Headquarters, and then just as suddenly oozing charm and plying him with drinks. One staff officer and weapons expert for whom Mannock retained genuine respect was Major Pearce. At Mannock's instigation the Wing's chief armament officer came up with a simple means of increasing the Vickers' rate of fire: the insertion of a washer on the front of the gun barrel. Mannock continued to work closely with Pearce, testing the latter's ideas on how to render weaponry more effective before passing on the results of their experiments to other squadrons. He introduced Pearce to McElroy, ensuring that the older man's technical advice would still be available to 40 Squadron after Mannock had gone home. For Pearce his 'disciple' was 'one of those lovable types who made one go full out to help him in every way'.[87] This was probably just as well, for Mannock's irascible behaviour towards anyone wearing the red or blue tabs of a staff officer extended as far up as Trenchard himself.[88]

At the end of 1917 Trenchard was waiting to return home and take up a new post as Chief of the Air Staff, in expectation of the RFC and RNAS's merger on 1 April 1918. However, in his last weeks in France he remained very much the hands-on general. Lewis was always a great admirer, but Mannock remained unimpressed, even though his MC and bar owed a good deal to Trenchard's personal interest in this most opinionated of pilots. By the time Trenchard arrived at 40 Squadron to see for himself what the problems were with the SE5a, Mannock's patience was exhausted. He had been horrified by his experience on Christmas Eve when an engine seizure at low altitude had made him an easy target for German 'Archie' as he glided westwards. Once safe over the British front line he was forced to land between the trenches, and then spent over three hours in a killing field, surrounded by rats and rotting corpses. The all-pervasive and unrelieved stench of putrefaction – he had no cigarettes with him and he was constantly sick – left the traumatized pilot screaming for somebody to come and rescue him. Exhausted and disgusted, once in the privacy of his own room he broke down.[89]

On the day of Trenchard's visit Mannock brought A Flight back to Bruay incandescent that his guns had jammed yet again. Still in the cockpit he lambasted the slow-firing and unreliable Vickers, insisting in no uncertain terms that if he had been flying his old Nieuport he would have destroyed at least two German aircraft. For all Tilney's efforts to act as if everything was normal, Trenchard had witnessed the incident, and overheard the ensuing conversation between an embarrassed CO and an irate flight commander who simply refused to keep quiet. The episode echoed Albert Ball's personal complaint to Trenchard the previous spring. All Ball's efforts to improve the original SE5's speed and rate of climb had proved fruitless, so he told the general to his face that the aircraft was a 'dud'. The *froideur* between Tilney and Mannock lasted only a few days once it transpired that Trenchard had been sufficiently taken aback as to order immediate action. As well as the staff officer sent to advise on the machine guns, an urgent investigation was ordered back home as to why the engines were so unreliable. Only then were the subcontractors' production difficulties revealed, and attempts made to improve quality control prior to a complete modification of the original design.[90]

Trenchard, and the commander of I Brigade RFC at Bruay, the youthful and precocious Gordon Shephard, admired Mannock because he was eager to maintain the offensive strategy throughout the winter, even if that meant flying greater distances in order to engage with the enemy. Lewis and MacLanachan, who by now was sharing command of A Flight, urged a more limited role of supporting ground troops, in order to conserve aircraft and pilots, and to become better acquainted with the SE5a. They believed the best German pilots had been withdrawn for the winter to act as instructors, while their aircraft were being re-equipped and refurbished in anticipation of a major aerial offensive in the spring.[91] As far as 40 Squadron was concerned this argument was increasingly academic as by mid-December regular patrols had virtually ceased, and the two most senior flight commanders, Keen and Lloyd, were given home postings. Attention at RFC Headquarters was focused further south on 56 Squadron. 56 had the élan, it had long since ironed out any engine problems, and perhaps most important of all, it had McCudden.[92] His reputation spread throughout the sector, and back across the Channel: Gwilym Lewis informed his parents in late December that, 'The expert out here now is McCudden...In about three days he brought down eight Huns, seven of which I believe fell this side. An unparalleled success.'[93] When McCudden visited Mannock at Bruay what most impressed MacLanachan was the ex-fitter's absolute faith in his aircraft – knowing his machine inside out he had no qualms about taking off with the engine cold. Ironically, when McCudden died in July 1918 it was as a result of engine failure in an aircraft he had no opportunity to become familiar with. He flew out from England to assume command of 60 Squadron in a new SE5a, but the machine had been fitted with the wrong carburettor. Landing at Aix-le-Château to seek directions, McCudden made a steep turn after take-off, and the Viper engine stalled – fatally.[94]

In *Five Years in the Royal Flying Corps* McCudden described how, drawing upon his ground crew experience, he markedly improved the overall performance of successive aircraft, and in particular the SE5a. McCudden's book was in itself a remarkable achievement, written in just over a month prior to his final departure for France in July 1918. Although he had only received an elementary education, the lengthy manuscript required minimal editing before being rushed in to print near the end of the war. The book suggests a staggering capacity for recall, unless McCudden kept detailed notes or even a diary throughout his service career.[95] In his preface Trenchard noted that, 'No detail, however small, connected with any branch of his work or any part of his machine was overlooked'.[96] The late author had himself insisted that:

> I always take a great personal interest in my machine, and I was rewarded by the knowledge that my machine was as fast and would climb as well as any other in the squadron... I am a stickler for detail in every respect, for in aerial fighting I am sure it is the detail that counts more than the actual fighting points.[97]

Like Mannock and 'McScotch', Bishop and Ball, McCudden spent days testing and resighting his guns, for which he was repeatedly 'chaffed' by the rest of his squadron. Only when he arrived at 56 Squadron did he find a gathering of like minds.[98] The arrival of pilots like Lloyd and Lewis, along with Mannock's growing success, ensured that by the end of 1917 40 Squadron was imbued with a similar professionalism. Nigel Steel and Peter Hart rightly identified McCudden as a 'player' not a 'gentleman', and the same could be said of Mick Mannock.[99]

McCudden's insight, and his capacity to survive for so long, depended upon technical expertise and the caution of a working-class lad who had learnt through harsh experience not to take for granted anyone or anything – man or machine. A Nieuport 17, and more especially an SE5a or a Sopwith Camel, were, by comparison with the aircraft that went to France in August 1914, complex machines. As has been seen, neither McCudden nor Mannock were in awe of the new technology, hence an eagerness to push their aircraft to the limit. In this respect they were by no means unique among RFC pilots, but their common experience exemplified how in twentieth-century war, where strategic/tactical initiative and technological innovation went hand in hand, training in applied science rather than the classics was all too often a prerequisite for survival. Both Mannock and McCudden attracted (Protestant) public school prejudice, but both overcame petty social snobbery. Having said that, the officers of 85 Squadron refused to serve under McCudden specifically because of his birth and background (and the spurious charge that, 'He gets Huns himself but he doesn't give anybody else a chance at them'). Strangely, the same squadron expressed no qualms about Mannock's appointment only a few months later.[100] Other than their age, the big difference between the two men was that McCudden, as might be expected of one in uniform throughout his formative years, remained polite and respectful to the very end (in *Winged Victory* he was described as being 'quiet and not disposed to riotousness' when obliged to socialize in the mess, and thus very different from the increasingly extrovert Mannock).[101] One reason why *Five Years in the Royal Flying Corps* required so little editing was that its author eschewed controversy, depicting everyone in authority in a positive light – every officer, without exception, was a gentleman.

MacLanachan and Mannock were posted back to Britain at the beginning of January 1918. McElroy, due for his first leave, returned home with them. A fourth travelling companion was McCudden, the party staying overnight in a Boulogne hotel before catching the ferry to Folkestone. This is the only documented occasion on which Ireland's three most successful fighter pilots congregated together, although there may well have been other occasions in France or at Bruton Street over the next six months.[102] Ironically, once the war was over all three were eclipsed in Ireland's national consciousness by an earlier member of 40 Squadron, Major Robert Gregory.

Gregory, who had flown with Mannock for the first two months he was in France, divided his life between Bohemia and the Big House, the Abbey Theatre and the Galway Blazers. As the adored offspring of Lady Gregory of Coole and a talented stage designer, the tiny but perfect Gregory was much admired by his mother's collaborator and protégé, W.B. Yeats – even though the feeling was never reciprocated. Already in his mid-thirties when he joined 40 Squadron in the autumn of 1916, his age, marksmanship and love affair with Paris, earned him the MC *and* the *Croix de Guerre*. Posted to command 66 Squadron in northern Italy, Gregory died in mysterious circumstances on a practice flight in January 1918, fuelling rumours that he had either passed out or been shot down by 'friendly fire'. Officially, he had been 'Killed in action'. His death prompted a grieving Yeats to write a lengthy eulogy in the *Observer* ('I have known no man so accomplished in so many ways... Leading his squadron in France or in Italy, mind and hand were at one, will and desire').[103] There followed four poems, of which two would rank among Yeats's finest work. Both 'In Memory of Major Robert Gregory' and 'An Irish Airman Foresees His Death' encouraged young writers to engage with flight and the cult of the aviator throughout the interwar period.[104] Unlike the high scoring McElroy, or the second-generation Mannock and McCudden, Robert Gregory became the embodiment of a strange and short-lived phenomenon, the noble Irish aviator. Elegized by the national poet, and yet dying for an already unfashionable and unpopular cause, Gregory's image and status remained curiously paradoxical. When in late 1920 Yeats contrasted the *chevalier* who shot down 'Some nineteen German planes, they say' with the Auxiliaries murdering his tenants at Kiltartan Cross, Lady Gregory insisted that the memory of her son should not be tarnished by what she saw as crude propaganda: 'Reprisals' was not published, as Yeats intended, in *The Times*; and the 'Irish Airman' never entered the nationalist pantheon.[105]

Gregory, a cultured and sensitive man, was remembered with respect and affection, but he had never stamped his personality on 40 Squadron in the manner of Mannock and later McElroy. The farewell dinner for Mick and 'McScotch' was an occasion for celebration and for serious reflection. The squadron's riotous Christmas party, orchestrated by the normally severe and abstemious Lewis, had appalled MacLanachan, hence this more sombre occasion.[106] Tilney paid tribute to both men, and in particular the commander of A flight. Later that night the CO recorded Mannock's departure in the squadron diary: 'His leadership and general ability will never be forgotten by those who had the good fortune to serve under him.' Urging his father – a governor of the LSE – to make early contact with this 'most arrogant socialist', Gwilym Lewis summed Mannock up as, 'one of finest personalities I have ever met. Very popular by all he met, and a regular hero in this squadron'. He was returning home with 21 victories to his credit, having downed a DFW on the eve of his departure. Even as the tender waited to take him to the coast he was airborne seeking a final, elusive kill.[107] When finally it was time to leave, 'the road was lined with cheering mechanics'.[108] Except when severely stressed Mannock had treated all the ground crew with generosity and respect, hence his popularity. A collectivist by instinct, and by gesture, he respected riggers and fitters as much as pilots and observers – constructing a successful squadron, let alone a more prosperous society, required teamwork and a genuine sense of ownership. Half a century later, Lewis looked back on Mannock's departure from Bruay:

> He left the squadron with 21 victories and his victories were good, he came on to form having been older than most of us and a more mature man. He had given great, deep thought to the fighting game and had reorientated his mental attitudes which was necessary for a top fighter pilot. He had got his confidence and he had thought out the way he was going to tackle things.[109]

Over the remaining seven months of his life Mannock was to enjoy remarkable success in disseminating his ideas across scout squadrons on how to minimize casualties while still relentlessly pursuing Trenchard's offensive strategy. By the time he returned to France the course of the war had changed dramatically: with the BEF in March–April 1918 forced to fall back on Amiens, the need to maintain Allied air superiority became absolutely essential.

5
74 and 85 Squadrons, 1918

Sojourn in Blighty, January–March 1918

Mannock spent the first three months of 1918 in England. The grounded fighter pilot eager to return to the fray is a predictable cliché, and yet it does seem that even while still on leave Mannock was fretting for action. Having spent some time in Birmingham staying with his mother and his sister, he arrived at Wellingborough in a 'nervous state'. Jim Eyles attributed Pat's moodiness to unease over the behaviour of both his mother and his elder sister, the one an alcoholic and the other an emotional casualty of a failed marriage. However, it quickly became clear that he was 'like a cat on hot bricks' mainly because the 'staff types' were denying him an early return to France. Like 'McScotch', Mannock believed the Germans would attack sooner rather than later: the war in the east was to all intents and purposes over, and it was imperative to launch a knock-out blow in the west before the full force of the Americans ensured an overwhelming Allied advantage in material and manpower. Expectation of Germany's renewed aerial offensive heightened his determination to get back into the war. Regularly catching the train down from Wellingborough, Mannock would reach Euston ready to tramp the long corridors of the War Office. Lobbying for an overseas posting, he received a unanimous reply, namely that he was too useful as an instructor to risk returning to combat. No doubt senior staff took seriously Leonard Tilney's recommendation that, in his own interest, Mannock be given an extended break from fighting. The Eyles family were left to pick up the pieces, nodding sympathetically as their caged warrior returned to Mill Road berating Whitehall's 'red-tape Brigade'.[1]

When not fretting about his future Mannock relaxed with one set of comrades in the Labour Club and a very different set in the RFC Club:

Eastfield or Berkeley Square, he was equally at home. Sometimes Jim Eyles would accompany his friend to London. Signed in as a guest at 13 Bruton Street, he would watch Pat Mannock, scourge of the bourgeoisie, become Mick Mannock, scourge of the Boche, once the front-line fraternity gathered in the bar to rag, rant and reminisce.[2] Eyles was with Mannock the weekend he confronted Major-General Sir David Henderson, the most senior figure in the RFC. By this time Mannock was stationed near Bromley in Kent, as a duty pilot at the Wireless Testing Park, Biggin Hill. With the RFC developing more sophisticated wireless communication with ground forces, its experimental work was crucial. This was of little consolation to Mannock, as for two days he flew an FE2 round in circles doing little more than acknowledging radio signals from the airfield below. He had told Eyles on 4 February that, 'I don't think I shall be able to remain satisfied at this delightful spot, and I am trying to get out to France as soon as I can.'[3] Forty-eight hours later the two men met at the RFC Club, where Mannock answered Henderson's casual enquiry as to his wellbeing with characteristic bluntness.

David Henderson was unusual among his fellow regular officers in being a graduate (twice) and the author of the first training manuals that acknowledged the importance of aerial reconnaissance. In order to test his ideas, in 1911 at the age of 49 he learnt to fly. Three years later he led the RFC to France, returning home in August 1915 to become Director-General of Military Aeronautics at the War Office. The following April he was appointed to the Air Board, a newly created supply and coordinating body foreshadowing the Air Ministry. When the Gotha attacks provoked a public clamour for greater security and reprisal bombing, Jan Smuts was asked to prepare two cabinet reports: on the state of the home defences, and the future of air power. Henderson was designated the Boer soldier-statesman's special adviser. He retained this role when, in August 1917, the War Cabinet authorized Smuts to chair the Air Organization Committee. This committee's task was nothing less than to create a unified, independent air force, complete with its own board and ministry. Enjoying a remarkable degree of autonomy out in the field, Trenchard had taken a very different view from Henderson. He felt the RFC needed time to consolidate after an extended period of upheaval, and that a high-profile third force would be subject to a much greater degree of political interference (he had resented having to redeploy his best scout squadrons in order to provide London with only nominal protection). Nevertheless, it was Trenchard not Henderson who became the first Chief of the Air Staff in January 1918. Henderson was again passed over when Trenchard resigned after less than three months, having found Lord Rothermere, press baron turned inaugural Air Minister, impossible to work with. Unable to accept the appointment of his former number two, the deeply unpopular Frederick Sykes, as the future RAF's second chief of staff, Henderson also submitted his resignation.[4] It was against this backdrop of Whitehall intriguing and clashing egos that Mannock had his brief but consequential exchange with the man whom many students of air war see as the real father of the Royal Air Force.

A man of intelligence and integrity, Henderson was by all accounts both popular and highly regarded. He was a key figure in the expansion of the RFC, and in particular the exponential growth in aircraft production after 1915. He preferred to work out of the public eye, which was one reason why the more widely known Trenchard became Chief of the Air Staff, but his immense achievements were recognized across the Corps. Mannock would have viewed Henderson in a very different light from Trenchard, but he was equally forthright when it came to expressing an opinion. When Henderson advised Mannock that he could expect to be on Home Establishment for at least another two months, the latter threatened to steal an aircraft and fly straight back to Bruay. Henderson pointed out that such action could only result in a court martial sentencing him to be shot. The immediate retort of 'Death is better than dishonour, sir!' clearly appealed to Henderson's sense of humour as he promised to see what could be done. Mannock immediately asked if in the meantime he could be given the opportunity to shoot down Gothas. Henderson said no, suggesting that the SE5a was not suited to night flying. Mannock again became angry, albeit not to the general's face: he probably remembered that the previous summer 56 Squadron had not encountered too many problems landing their SE5s in the dark.[5]

Henderson took immediate action, and Mannock was swiftly 'restored to the Club fireside, pending disposal'.[6] He never returned to Biggin Hill, and a few days later wrote to Eyles with news of a posting to 74 Training Squadron at London Colney, an airfield in Hertfordshire. 74 was actually at quite an advanced stage of training. It might better be described as a nucleus front-line squadron which was in its working-up period prior to the arrival of new equipment, followed by deployment overseas.[7] The squadron comprised seven instructors and 32 pupils, but the final complement would be only 20. With less than four weeks until 74 became an active service squadron, and instructors like Mannock automatic selections, it's not surprising to discover that there was keen competition for inclusion. We know this because none was keener to be selected than an ex-observer from west Wales, Ira Jones, predictably known to all and sundry as 'Taffy'. Jones was born into a poor Welsh-speaking family which laboured on the land around St Clears, a tiny community west of Carmarthen. Scarred by village gossip that he was illegitimate, and afflicted by a stammer, Jones was unusually small, even by the standards of an under-nourished Edwardian working class.[8] In 1918 he was 22, and had already spent around 18 months on the Western Front, first as ground crew and then as an observer. Long hours apprehensive and exposed above Arras and Albert had earned him a well-deserved Military Medal. By dint of his background Jones had much in common with Mannock, not least the fact that, prior to joining the RFC in June 1915, he was a telegraphist in London.[9] Jones was clearly bright, the Cana Chapel and School fostering a natural writer. Between the wars he compensated for a flagging service career by earning the respect of Fleet Street's features editors and Twickenham's press box. But back in 1918 he was still keeping his thoughts to himself, via his voluminous diary entries, replicated in *Tiger Squadron*. Jones chronicled 74's 12-month transition from an advanced training unit of trench veterans, hardy colonials and retread observers into the battle-hardened 'Tiger Squadron', boasting over 200 victories for the loss of only 15 dead and five captured. Other than the squadron diary, and each pilot's combat reports, Ira Jones is therefore the main source of information about 74, even though his entries read suspiciously as if they have been rewritten for purposes of publication.[10] In *King of Air Fighters* Jones quoted copiously from his diary, but with slight changes from the original text. He attributed these entries not to himself, but to the fictional 'Lieutenant James VanIra, a Welsh-South African, the spare pilot'. In 1934 Jones was still a serving officer – he retired from the RAF for the first time two years later – and by keeping a diary when on active service in France he had theoretically been in breach of regulations.[11]

Jones clearly felt about 'Captain Edward Mannock, the greatest patrol leader of any fighter force in World War I' the same way that William MacLanachan did.[12] But a close reading of *Fighter Pilot* reveals a readiness to acknowledge Mannock's faults and expose flaws in his thinking; indeed 'McScotch' claimed to have been a major influence when it came to putting new ideas into practice. Jones was far more deferential and reverential, even though in many respects *King of Air Fighters* and *Tiger Squadron* are more informative (the former not least because of Jim Eyles's collaboration). Recording his first impression of Mannock, Jones set the tone for both his biography and his squadron history:

His tall, lean figure; his weather-beaten face with its deep-set Celtic blue eyes; his unruly dark-brown hair; his modesty in dress [stained and faded RE tunic] and manner appealed to me, and immediately, like all the other pupils, I came under his spell. He had a dominating personality, which radiated itself on all those around. Whatever he did or said compelled attention. It was obvious that he was a born leader of men.[13]

He certainly wasn't a born flying instructor, leaving the finer points of landing or taking off in an SE5a to the technically proficient. Mannock had been assigned to 74 because he had recent combat experience, and was therefore familiar with the Albatros and the Fokker triplane, and above all because he was a tactician.

In *King of Air Fighters* Jones justified his claim that Mannock was 'the first Allied airman to realise the supreme importance of applying tactics to formation fighting'. The Welshman attended Mannock's lectures, both before and after embarking for France. Copious notes formed the basis for Jones's résumé of Mannock's ideas 16 years later. The new instructor was an effective communicator and an intuitive motivator. His daily talks were genuinely interactive, with his pupils set problems and invited to solve them. Mannock would then comment on the answers, criticizing and advising where appropriate. General principles of airfighting were complemented by detailed guidance on how to respond to a range of different scenarios. His advice was rooted firmly in common sense and a firm insistence on avoiding heroics when faced with a superior enemy ('don't ever attempt to dog-fight a Triplane on anything like equal terms... run for home like hell, kicking your rudder hard from side to side in order to make the shooting more difficult for the enemy, but – still praying hard'). Not surprisingly, great emphasis was placed upon individual responsibility, particularly when it came to the sighting and loading of machine guns: pilots needed to be familiar with the specification and performance of their equipment, and their scientific and technical knowledge had to embrace meteorology as well as aerodynamics and ballistics. They needed to know the weaknesses of their own machines, and those of the enemy, especially those blind spots most advantageous to the attacker. New pilots who had not previously served as observers had to train their eyesight to spot small objects at long range, and rapidly establish distance, numbers and identity. All apprentice pilots were ordered not to waste time practising fancy stunts but to focus on simple, life-saving manoeuvres, most notably the quick turn: there must be no straight flying unless firing, and above all, no offering the enemy an easy target by diving away from him.[14]

Mannock insisted that, other than in exceptional circumstances, an attack should only be initiated when a patrol enjoyed a clear numerical advantage. The German fighters' manoeuvrability and fast rate of climb necessitated superior numbers and an element of surprise. The key was always to attack from a superior position, hidden by cloud cover or the glare of the sun, and ideally from the *east*, i.e. approaching enemy aircraft from behind their own lines. Drawing upon his experience with 40 Squadron the previous autumn, Mannock made the case for high altitude patrols, forever emphasizing the importance of attacking from above: diving at maximum speed, a confident and well-practised pilot could delay fire until the last possible opportunity, ideally well within one hundred yards of his chosen target. Once engaged in pursuit of an enemy aircraft a pilot had to resist any temptation to follow his quarry down to ground level: flying eastwards in hot pursuit just above the German trenches would leave him vulnerable to machine gun fire.[15]

Mannock's mantra, chanted at the beginning and end of every lecture was: 'Gentlemen, always above; seldom on the same level; never underneath.' 74's most popular instructor wanted consistency, but he also urged flexibility and personal responsibility. Members of the flight were expected to follow their leader in formation only until the latter signalled the conditions were right to attack. Mannock believed that diving in formation risked collision, while spearhead attacks led by the flight commander were deemed equally dangerous. Airfighting – just like socialist solidarity – had to accommodate individual flair and endeavour, and yet the circumstances of 1917–18 dictated that it had to be a collaborative effort. Unlike Ball or Bishop, and despite his single-mindedness and idiosyncrasies, Mannock was a team player. In 74 Squadron all advanced training was built around forging a keen sense of identity and camaraderie, and many of the team-building techniques, although crude, would be familiar today. He used humour a great deal to get his ideas across, but he could also be a strict disciplinarian as and when required. Once in France, anyone who broke formation too early, thus endangering his comrades, would face the flight commander's wrath back at base, or even, on at least one occasion, be given a warning burst of 'friendly fire'. He wasn't averse to requesting a transfer out of the flight for anyone reluctant to toe the line. Debriefings were genuinely diagnostic, with Mannock carefully explaining what he had been seeking to achieve, and who had performed well or poorly, and why. He only lost a handful of pilots in the flights he commanded in 40 and 74 Squadrons, and as we shall see, replacements were protected for extended periods. New arrivals were given an early boost to their self-confidence by sharing and then claiming one of Mannock's 'kills' – a practice that he initiated when tutoring McElroy, but which ultimately would cost him his life.[16]

In early March, 74 was redesignated an active service squadron, and placed under the command of Major Alan Dore, recently returned from France where he had been on active duty for over two years.[17] Dore was actually in urgent need of a rest, and his command lasted only two weeks, during which time the squadron prepared for service overseas at No. 1 School of Aerial Fighting, Ayr. With instructors of the calibre of McCudden, those novices who had survived the selection process were given individual tuition by some of the finest fighter pilots currently out of the front line. 74's brief stay in Scotland was a rare opportunity for McCudden and Mannock to work together, both conveying a common approach to the war in the air, but both well aware that even the most sophisticated training was no match for direct experience. Each of the three flights that made up 74 Squadron contained at least two experienced fliers, while most of the newly trained pilots had already seen some form of action, in either France or east Africa. The disproportionate number of volunteers from the Empire – three out of seven in Mannock's A flight – meant that few of the squadron were direct from school or university. Given that 74 included an American and an alleged distant cousin of von Richthofen, it is scarcely surprising that Mannock's background no longer appeared unusual.[18] His presence in the mess was especially welcome to Dore's successor, Major Keith 'Grid' Caldwell MC.

Caldwell was a well-built, swarthy and extremely tough New Zealander from Waikato, whose nickname came from his habit of calling all aircraft 'grids'. He had flown with both Ball and Bishop, and his fame rested on a seemingly insatiable appetite for aerial combat. Flying with 60 Squadron Caldwell had acquired a reputation for always being able to sniff out the enemy. He was aggressive and ruthless, and if he had been a better shot then his tally of victories might well have matched those of Bishop and Mannock.[19] Caldwell had met the latter when visiting 40 Squadron, quickly concluding that here was a man after his own heart. His initial impression was confirmed when he arrived at London Colney on 8 March to assume command: 'I had every confidence that Mannock was just the right type to inject the right sort of attitude into the chaps and give them a good chance when we got up against the Hun.'[20] Caldwell liked nothing better than to fight Germans, but he had more in common with his new flight commander than with his distinguished former comrades: the first time he addressed the squadron he placed great emphasis upon teamwork and mutual support. He had a relaxed Antipodean view of rules and regulations, but was a strict disciplinarian when it came to flying. In this respect Mannock and Caldwell were of like mind, both men agreeing that squadron morale was paramount. One of the two young Canadians in A flight, 'Clem' Clements, later paid tribute to the quality of leadership in 74 Squadron:

> We developed into a family really, Grid and Mick saw to that, an efficient, happy team...On the rare occasions that gloom did settle in the mess, Mick was just the man to handle it. He didn't give a hoot how he did it, as long as the men ended up happy and morale was maintained. He was always the life and soul of the party, although this never interfered with our respect for his authority.[21]

Caldwell's arrival signalled the re-equipment of the squadron. Nineteen new aircraft were delivered, tested and made ready for active service. During the day pilots practised their gunnery and formation flying, and in the evening would be briefed by Mannock on what to expect the first time they crossed the lines. On 25 March the squadron was given five days' notice of deployment overseas. Waiting at an airfield in Essex for confirmation that the squadron's equipment and baggage had arrived in France, Mannock secured permission to fly to Wellingborough. Surprisingly, he invited the youngest member of his flight to join him, and upon their arrival a car whisked them to 183 Mill Road. Clements was astonished to discover that Mannock's standing in the community was not just because of his exploits as an aviator. Here was someone whose prewar immersion in local politics meant that he had a remarkably wide circle of friends and associates: 'It seemed as if the whole of Wellingborough had turned out to greet him.' When they dined with Jim and Mabel Eyles at the Hind Hotel in the centre of town, it became clear that Pat/Mick was on more than nodding terms with the local great and the good.[22] Not only was he a war hero, but he and his dining companions were the key figures in a constituency party only eight months away from electoral victory. The presence of a young Canadian flier added glamour to the occasion, further subverting the notion of Labour as a dour and unpatriotic agent of the politics of envy.

Veteran flight commander

On the morning of 30 March, 74 Squadron flew south to Lympne, and then across the Channel to Calais. After a day at St Omer all three flights returned to the coast for gunnery practice off Dunkirk, prior to commencing regular patrols over the line. A further week was spent away from the action, before Caldwell secured the squadron's transfer to XI Wing, based at Clairmarais, not far from St Omer.[23] While being 'chivvied & chased from one aerodrome to another', Mannock found time to visit Bruay, where his old mess mates gave him a rousing reception. Bemoaning BEF bureaucracy and the atrocious weather, he pointed out that he had been a fortnight in France and, 'I haven't yet seen a Hun! My guns are rusting through lack of exercise.'[24]

By the time 74 belatedly established itself as a front-line squadron the RFC was no more: the Corps had officially merged with the RNAS on 1 April to form the Royal Air Force. The launch of a third service could scarcely have taken place at a less propitious time. Notwithstanding the politicking and infighting within the Air Staff and the Ministry, in France the RAF – like the rest of the British Expeditionary Force – was facing a crisis of unprecedented dimension. On 21 March 1918 no less than 76 German divisions, across a 50-mile front, had launched 'Operation Michael'. Fresh troops (from the east) and fresh tactics enabled the Germans to penetrate 40 miles behind the southern half of the British front line. The 15 divisions of the Fifth Army, the 14 divisions of the Third, and the eight divisions that made up GHQ Reserve, were all forced to retreat, often in total confusion. Over 1000 square miles of ground was captured before the initial Somme offensive petered out on 5 April, with the Fifth and Fourth Armies at last holding a line ten miles east of Amiens. Between 9 and 12 April, Ludendorff took advantage of poor visibility and restricted flying to launch a fresh attack further north, seeking to capture the important communications centre of Hazebrouck, and thus punch a way through to the Channel. Having failed to provide adequate air cover for the initial ground offensive, in Flanders the *jastas* adopted more aggressive tactics once the weather improved. Again, although stretched to the limit, the RAF maintained the same level of air superiority that had proved such a key factor in preventing the retreat to Amiens from becoming a rout. The RAF had the advantage of flying out of undamaged airfields, whereas the German Air Service found difficulty in establishing advanced landing grounds. In consequence, German pilots were having to fly ever greater distances before reaching the front line – unlike their opponents who could respond rapidly to intelligence reports of fresh troop movements, and strafe and bomb accordingly. Not that the British did not sustain serious losses: by the time Ludendorff's northern offensive ground to a halt on 29 April, the RFC/RAF had lost over 1000 aircraft in five weeks. With sorties flown continuously from dusk to dawn, fatigue was bound to take its toll. Some squadrons had suffered up to 30 per cent casualties daily, the equivalent of a complete change of personnel every four days.[25]

74 Squadron entered the battle for the first time on 12 April, appropriately the day Haig issued his famous injunction that, 'With our backs to the wall, and believing in the justice of our cause, each one of us must fight on to the end.' In a moment of genuine high drama Caldwell gathered all his pilots to the mess and read out the order of the day. A sombre squadron, officers and men, then assembled outside, and cheered the inaugural patrol until it finally disappeared eastwards.[26] Mannock recorded 74's first victory, and in the afternoon shot down a second Albatros. Amid great celebration he ensured the whole of A flight received due recognition, albeit berating Clements for getting lost. The latter refused to be intimidated, and 'After this I never had any trouble and flew as Mannock's wing man. One could say I became his shadow.' Jones, in C flight, received a similar tongue lashing for chasing a two-seater at low level over enemy lines.[27]

74's first day of action set a precedent for Mannock's handling of his flight, and indeed the whole squadron. Indiscipline would be firmly dealt with, and any further indiscretion might well be met with a disproportionate response. Most pilots accepted this as the price of flying with an inspirational air fighter whose first priority was to protect his men. Mannock was an aggressive leader, but everyone on patrol had a designated task, not least the need to scour the skies while he focused on spotting and stalking the enemy.[28] So long as they fulfilled their responsibilities then each member of the flight would not only enjoy repeated praise and encouragement, but they were given every opportunity to share in a victory. Caldwell insisted that his senior flight commander was selfless, and when appropriate would ensure a new pilot could claim his first kill: 'He would not hand these out piecemeal; he had to feel that the pilot was worth the encouragement.'[29] Thus, on 30 April Lieutenant Dolan, an ex-artilleryman Jones considered 'Mick's protégé', shot down a two-seater after 'Mannock deliberately killed the gunner'. Once blooded, Dolan went on to destroy eight enemy aircraft before he himself was shot down a month later.[30] Similarly, by mid-May Mannock felt something had to be done about the youngest member of A flight, 'Swazi' Howe, a slight 17-year-old South African seemingly incapable of shooting straight. Day after day he took Howe out 'on a private war to see if he could help him get a Hun'. On 18 May Mannock blooded the tiny South African by convincing him that he had shot down an Albatros. Almost certainly Howe's contribution had been secondary, if only because in a dogfight where the two men were easily outnumbered Mannock's aggression, tight flying and close shooting were crucial to survival. He had linked up with a flight of Sopwith Camels at the same time that he spotted six Albatros, and had immediately signalled an attack. Howe and his flight commander dived, but in the ensuing dogfight quickly established that they were alone. The Camels, part of an ex-RNAS squadron whose Australian CO was unashamedly hostile towards former RFC units, had not actually joined the attack. 'Lucky to get away with whole skins', both the SE5a pilots crash landed, with their machines badly shot up. An incandescent Mannock insisted Caldwell complain directly to Pierre van Ryneveld, the South African colonel commanding XI Wing. When pressed on the matter, the Camels' squadron commander claimed that just as Howe and Mannock dived his patrol spotted, and subsequently intercepted, a second enemy formation. A sceptical 74 strongly suspected the Camel flight had funked the fight. Such a suspicion could never be proved, but the incident confirmed that the fledgling RAF was still plagued by lingering inter-service rivalry.[31]

Mannock kept up the routine of thorough post-patrol debriefings, complemented by talks to the whole squadron. The emphasis was on teamwork and keeping morale high, with frequent inter-squadron rugby matches to reinforce a growing sense of camaraderie. Caldwell provided the bloodcurdling exhortations to take the battle to the enemy, always leading from the front, whether in the air or on the rugby field.[32] Like Tilney, he paid great attention to detail, convinced that a well-run and well-stocked mess was crucial to squadron morale. For all the improvisation and scrounging, mess life at Clairmarais followed a familiar prewar model. During the day officers relaxed by listening to the gramophone, and reading the previous week's issues of *The Times* or the *Tatler*. (Mannock had little interest in magazines, and it was not unusual to find him seated in a leather armchair reading Shakespeare or Tennyson.) In the evening officers would converse over sherry before white-jacketed orderlies served a four-course dinner. The following day's operations were finalized over dessert, with the orderly officer interrupting his meal to take notes. More formal functions would end with a morale-building address from A flight's commander, who by all accounts was a first-class after-dinner speaker, skilfully mixing gravitas and humour. Although Caldwell went to great lengths to secure decent furnishings and fresh food, he ran a tight ship: wine and port were reserved for special occasions, as were mess games, and anyone on the dawn patrol was expected to retire early. He was always up to see off the first patrol, and in late afternoon led 'a full squadron show' over the German front line, albeit rarely provoking the enemy into taking up 74's mass challenge.[33] 'Grid' Caldwell was unusual in enjoying two nicknames. 'Marshall Ney', although a pleasingly Napoleonic sobriquet for any squadron commander, was a double-edged compliment: it acknowledged his bravery, but signalled that his indifference to tactics and his cavalier approach to air fighting were not always appreciated. Jones later recalled that, 'Some of the patrols he led were nightmares, in which we were as frightened as the enemy.' For all his 'amazing flair for air fighting' his gunnery was so poor that he shot down remarkably few aircraft given the amount of time he spent in the air. At XI Wing's Headquarters Pierre van Ryneveld's admiration for Caldwell was clearly qualified as, when the 'Tiger Squadron' began gathering plaudits and decorations, its CO was denied even the DSO.[34]

While Caldwell provided the courage, Mannock kept the pilots on their toes when it came to tactics:

> Mick made us imagine a situation and then asked for a quick answer. In this way we started to 'think' and 'eat' air fighting... Practice in the air and his training sessions made us react automatically and instinctively to any situation which arose in a fight.[35]

If pilots had particular problems, not least their inability to hit a moving target, then Mannock provided technical advice on sighting and calculation of speed and direction. He passed on lessons learnt in his first weeks with 40 Squadron, not least the importance of frequent target practice even when engaged in regular combat.[36]

The relentless attention to detail, let alone the need to appear constantly at one with the world, was already taking its toll. Only a day after shooting down his two Albatros, Mannock was already intimating to Jim Eyles that the demands of front-line service were no less than when he was serving his apprenticeship with 40 Squadron.[37] Still haunted by the prospect of going down in a 'flamer', Mannock's strategy for dealing with this fear was black humour. Like Caldwell, Mannock tried to make light of every pilot's worst nightmare by cracking tasteless jokes about 'sizzlers'. The laughing stopped on 21 April when 74's first casualty, Lieutenant Begbie, was shot down in flames. Unfamiliar with Trenchard's insistence on minimal mourning and 'no empty chairs', those pilots new to France were horrified by Caldwell's insistence on a full mess dinner and the usual daft games. The CO explained the paramount importance of maintaining squadron morale as casualties mounted, and henceforth members of 74 grieved only in private. The same evening Mannock gleefully delivered the news that von Richthofen was dead, scorning any notion of a toast to this supposedly gallant foe. Drinking to Begbie's memory, he hoped the 'Red Baron' had died a similarly horrific death. Any hint of sentiment or honour was deemed pathetic, childish and naive: beyond memorial services and *Morning Post* obituaries, chivalry had no place in modern warfare. The respect shown to von Richthofen by a number of other squadrons (including burial with full military honours) was singularly absent from the mess at Clairmarais.[38] Absent too at Bruay, as confirmed by Gwilym Lewis: 'Never admired him very much as he was such a boaster. I don't think the Huns were very fond of him either.' When Mannock next visited 40, there could only be one topic of conversation, with Lewis and the rest of the squadron 'quite bucked that Richthofen is under the soil'.[39]

In early May, Caldwell and Mannock's public personae were tested to the extreme. Despite 15 confirmed kills by the end of its first month in France, the squadron suddenly lost three pilots in 24 hours, one due to a fatal stunt over the airfield. To make matters worse, Dolan, with whom Mannock had already forged a ferocious and profitable partnership, was shot down over Wulverghem on 12 May. Mannock's three victories in that same dogfight was no consolation for the loss of Dolan, 'a very full-out guy, and very popular...a bosom pal of Grid and Mick'. As usual on such occasions the squadron dined and partied, with Mannock the master of revels, but afterwards he retreated to his hut and 'wept like a child'.[40]

It would not be an exaggeration to say that Dolan's death had a profound psychological effect on Mannock, matched only by the loss of McCudden two months later. He became more irascible, losing his temper while on patrol, and more publicly but less dangerously when back on the ground. An 18-year-old replacement pilot called Sifton lasted only four days before he broke down in front of Caldwell and Mannock: 'Grid and Mick were speechless. Mick had the fellow's wings torn from his tunic.' The sobbing Sifton was despatched back to England within an hour.[41] Although Mannock had a reputation for being extremely supportive of young, green pilots who were chronically under-trained, and thus extremely vulnerable, he could scarcely contain his fury when one blustering arrival tried to disguise his inability to control an aircraft as powerful as an SE5a – another hapless adolescent was dutifully sent on his way.[42] At the same time, Mannock's fear of being shot down in flames was becoming more and more of an obsession:

He was particularly fond of cornering a new boy and describing for him the fall of a burning machine. This was done humorously but in a highly detailed manner. He left none of the gory facts out. If the new boy joined in the laughter, Mick would declare him to be OK.[43]

It was this treatment which had reduced Sifton to a quivering wreck. Even the starry-eyed Ira Jones recognized that 'Mick, who gets most of his Huns in flames, is getting very peculiar over the business.' At least six of Mannock's victims in May 1918 burnt to death, and on each occasion when he landed there would be semi-hysterical celebration and a gory inquest: 'Having finished in a frenzy of fiendish glee, he will turn to one of us and say, laughing: "That's what will happen to you on the next patrol, my lad." And we all roar with laughter.' The mood would soon change, with Mannock now gloomy and reflective, leading Jones to conclude, 'he is getting obsessed with this form of death. It is getting on his nerves.' Later, in *King of Air Fighters*, Jones recalled how Mannock kept insisting, 'I'll blow my brains out rather than go down roasting', and throughout his time in France he made frequent references to the fateful day he would need his Colt pistol. As well as the constant reference to 'flamers', a further sign of strain was that he started crashing aircraft on a regular basis, including three in one week.[44]

Mannock's ferocious treatment of any novice pilot who failed to live up to his expectations was, to be fair, unusual. Clements noted that, 'He had no use for slackers, but until he found them hopeless he gave them endless encouragement.'[45] The quality of pilots posted to 74 Squadron was unusually high as Caldwell, a former instructor at the Central Flying School, kept in close contact with the supervisory staff. 'Zulu' Lloyd, an admirer of both Caldwell and Mannock, screened CFS graduates in an attempt to ensure 74 received only the best. Anyone fresh from England assigned to A flight was given at least a week's intensive training under his commander's direct supervision before he flew anywhere near the front line, and even then he was under orders not to take any unnecessary risks.[46] Nor was Mannock's interest in inducting new pilots restricted to 74 Squadron. His ideas on how to ensure fresh arrivals survived beyond their first few weeks in France had spread far beyond Clairmarais. Similarly, his reputation as a proven tactical innovator had encouraged other squadrons to adopt clear guidelines on fighting in formation. Just as on the ground training and tactics were belatedly becoming uniform across the BEF, so above the trenches Mannock's collectivist approach to airfighting was increasingly the norm. The new orthodoxy eschewed individual endeavour, focusing on the need for teamwork and genuine fighting units. Yet ironically it also assumed a much greater degree of individual responsibility, with consequent rewards: by reducing the expectation that flight commanders alone could shoot down enemy machines, supporting pilots were given a much greater chance of securing a kill. Flight commanders still had to provide a clear lead, not least when initiating an attack from above, but, unlike in von Richthofen's *Jagdgeschwader*, they could not expect the rest of the patrol to be little more than a Praetorian Guard. Here was a realistic approach to airfighting, acknowledging that once aloft, in the words of 45 Squadron's Norman Macmillan, 'spiritually and emotionally we were shut in – we were self-contained individuals':

> Everything had to be the thought and action on the part of the one individual; he was entirely and inseparably alone. Even when he looked outside and saw the machines beside him, even then he still felt shut in responsible almost entirely for himself and every action. In combat it was the individual in the machine who had to make the decisions – not the man outside. It might be that the leader led others into action but once action was joined every man had to fend for himself.[47]

In order to maximize its offensive potential, and exploit fully its quantitative advantage, the RAF needed to ensure tactical consistency across its front-line squadrons. Thus, like Ball before him, Mannock's periodic visits to other squadrons became more systematic, with XI Wing arranging for him to give regular lectures on the new tactics. By the late spring of 1918, Mannock's reputation as a tactician was perhaps matched only by another SE5a pilot, Sholto Douglas, whose command of 84 gave him every opportunity to develop the ideal squadron formation.[48]

As if to revenge Dolan, Mannock sought every opportunity to engage with the enemy, flying his normal patrols, and then in tandem with Caldwell penetrating deep into enemy-held territory – the terrain was familiar from his tenure with 40 Squadron, but for the moment at least it was overrun by the Germans. On the ground, the late spring of 1918 saw Ludendorff launch a succession of limited attacks to the south, prior to a renewed summer offensive in Flanders. The Allies, especially now the French, focused upon containing these attacks. At the same time preparations advanced apace for an early counter-attack across a wide front. In the air, Ernst Udet had replaced von Richthofen as the *Jagdstaffel* talisman, and, as if to rub salt into the wounds of aggrieved and quietly seething RAF pilots, in late June a parachute saved his life.[49] While Udet was now the subject of mess gossip on one side of the line, on the other the Germans were mistakenly giving the same attention to McCudden. Mannock's total of 32 confirmed and unconfirmed victories since 12 April, of which half dated from after Dolan's death, led enemy intelligence to conclude that McCudden VC was back in France as a squadron commander. XI Wing fed this information back to 74 Squadron as an incentive to maintain its present high rate of attrition.[50] It was probably Mannock's growing predilection to stalk alone or in partnership that misled the Germans into assuming McCudden was back. Yet, as Jones's diary confirmed, the majority of his victories were still secured by leading A flight into attack, 'using dive-and-zoom tactics' and firing at exceptionally close quarters.[51] While noting Mannock's importance as a tactician, Caldwell attributed his consistency to: 'extraordinarily fine deflection shooting once he was engaged. In an air fight most people try to get behind the other man to get an easier shot & where you cannot be shot at (in the case of fighter v fighter), but Mannock was able to hit them at an angle.' He flew close, conscious of the enemy's blind spots, and as an attacker he was persistent – it was extremely difficult for a chosen target to shake him off, and in this he set an example to the rest of the squadron.[52]

Further boosts to squadron morale resulted from Ira Jones being awarded the MC, and in late May Mannock securing a DSO and bar in rapid succession. Taken together, Mannock's awards constituted an early acknowledgement of both his courage and his leadership since returning to France. The second citation focused on the recent destruction of, 'Eight machines in five days – a fine feat of marksmanship and determination to get to close quarters.' The *London Gazette* declared that, 'As a Patrol Leader he is unequalled,' but by the time it did so, the pilot in question had already been dead for seven weeks.[53] Unlike Ball and Bishop, and even McCudden, in his lifetime Mannock never gained the reputation at home which he enjoyed within the BEF. In France even Second Army's General Sir Herbert Plumer was aware of Mannock's exploits, insisting on meeting him during a surprise visit to Clairmarais. The monocled general in his Saville Row uniform and the dishevelled, dirty aviator in sidcot and muffler held a brief conversation, which left Plumer both bemused and amused. Mannock on the other hand realized that he had just been congratulated on receipt of a medal he knew nothing about. It took a telephone inquiry from Caldwell to Wing HQ to ascertain that Mannock had been recommended for the DSO.[54] Word spread fast, as the same day as Plumer's visit Gwilym Lewis was able to phone his friend and congratulate him: 'he told me he has now got 41

Huns, so he is probably at the head of the list of people at present out here.' Lewis noted that, a fortnight after Dolan's death, Mannock still keenly felt the loss of his 'fighting partner'.[55]

74's party to celebrate Mannock's medal included several pilots from 85 Squadron, who had arrived in France only two days previously. Notable by his absence was the squadron commander, Major 'Billy' Bishop. The Canadian ace already had 47 victories to his name, and the previous year had been awarded the VC. Since then he had returned home to encourage a fresh wave of volunteers, and – like McCudden – he had written a book. As we have seen, Bishop was a solitary figure, both in the air and on the ground, but he had taken personal responsibility for choosing his squadron, few of whom were British but all of whom were experienced scout pilots. In the event Bishop was in France for only a month before being recalled to help establish the Canadian Flying Corps. In that time he officially destroyed 25 enemy aircraft, most of them while flying alone. He had little if no contact with his closest rival, but his pilots certainly did. Mannock provided 85 with leadership and inspiration well before he assumed command on 5 July.[56]

In Bishop's absence 85 Squadron was usually led by K. K. 'Nigger' Horn, a South African captain who had flown with Caldwell in 60 Squadron, during which time he had been awarded the MC. On the evening of 29 May Horn and Mannock combined their flights, the latter assuming command. Only minutes earlier he had secured his first victory of the day. Over Lille the combined force encountered a similar number of Albatros scouts, and in a dogfight of unusual scale and intensity Mannock destroyed a further two aircraft. A similar confrontation three days later saw Mannock again claim three enemy fighters, having shot down a further Pfalz scout the previous evening. These victories were not without cost, both sides demonstrating unprecedented levels of ruthlessness, determination and ferocity. Thus, in the dogfight on 1 June Mannock's success was counterbalanced by the loss of W. E. Cairns, commander of C flight and a popular member of the mess.[57]

Although Jones and MacLanachan went to great lengths to suggest Mannock had little interest in his tally of kills, insisting that personal ambition counted for little when compared with pride in the squadron, his letters to the Eyleses reveal keen rivalry with McCudden. On 7 June Pat informed the 'Comrades' that, although officially accredited with 47 victories, he calculated the real figure was 51: 'I have only to beat Mac and old Richthofen now. I hope I shall do it.'[58] The same day, writing as 'the well-known hatchet-faced birdman', he conveyed a similar message to one of his old instructors ('Not so bad for an old chattering wreck like me what!').[59] The same week Mannock paid one of his regular visits to 40 Squadron, after which Gwilym Lewis wrote to his parents:

> He [Mannock] has now got a bar to his DSO and by bringing down three the night before has now 51 Huns to his credit. James McCudden was the leader on the British side before with 52, so you see you know someone pretty important![60]

When in London at the start of the year Mannock had accepted an invitation to visit 'St David's', the Lewis's large family home in Hampstead. Hugh Lewis's Fabian connections, let alone his enthusiasm for flying, must have been a great incentive – here was a man with the *New Statesman* in one hand, and a Royal Aero Club Certificate in the other. One consequence of tea in Templewood Avenue was that Mannock began to correspond with Gwilym's sister, Mary, a student in Bangor at the University College of North Wales.[61] In the final two years of the war Mary Lewis wrote regularly to a number of soldiers stationed on the Western Front, including Edgar McLaughlin, a lieutenant in the Australian Squadron which shared Clairmarais with 74 ('old Mick or rather Mannock is on our 'drome... we often have dinner together, he is always coming back with two or more huns').[62] Early formality was soon dropped, and the letters between Mary and Mick are relaxed and intimate. Nevertheless, the latter found it hard on occasions not to patronize – witness supposed terms of endearment such as 'my dear child', and his blunt dismissal of Lewis's belief in the spiritually uplifting purpose of music and poetry ('spare me the Sappho'). To be fair, for all Mannock's accomplishments as a parlour virtuoso, seen from the perspective of the Western Front it was hard not to scoff at any suggestion as to music's potential for nourishing 'the soul, character, outlook, [and] sense of justice'.[63] Presumably von Richthofen regularly listened to Beethoven – and Wagner. As we shall see, Mannock's later letters to Mary Lewis are revealing, even disturbing, not least in his efforts to reconcile the ideals of the Second International with a deep and remorseless loathing for the Germans.

Mannock's antipathy towards the enemy was further fuelled by the even greater risks scout squadrons were now expected to take. By deliberately avoiding engagement with the RAF over the front line, the German Air Service enticed SE5a patrols to fly ever deeper into enemy-held territory.[64] The more aggressive flight commanders cheerfully took up the challenge, albeit with the risk of running out of fuel if the flight home was slowed down by seasonal strong winds. With renewed

German pressure around Ypres, Mannock spent his final fortnight with 74 Squadron fighting in the flat Flanders fenland to the south of the Salient. His main victims were Albatros or Pfalz scouts, but on 6 June east of Ypres he downed his first Fokker DVII. With leave looming, and McCudden's return to France imminent, Mannock was eager to maximize his score before returning home in mid-June. At the same time, the consistent success of his 'dive-and-zoom tactics' meant that he was beginning to take unnecessary risks: when his patrol ambushed 18 Pfalz scouts over Zillebecke Lake on 16 June, they were outnumbered three to one. Mannock himself shot down two enemy aircraft, but he was the only member of the flight to enjoy any success, presumably because the others were too busy trying to keep alive.[65] Jones flew two patrols with Mannock the day before he was due to go on leave: a Halberstadt two-seater was coolly and remorselessly despatched, and a Fokker set up as an easy target for one of 74's original instructors, 'Dad' Roxburgh-Smith, the oldest pilot in the squadron. Unsettled by a fortnight at home, Jones thankfully recorded that, 'My blood lust had been re-awakened. My confidence was returning. Good old Mick!'[66] Nevertheless, in his diary the following day he again noted that Mannock's

> ...nerves are very much on edge. It is easy to spot when a pilot is getting nervy. He becomes very talkative and restless. When I arrived in the mess this morning, Mick's greeting was: 'Are you ready to die for your country, Taffy? Will you have it in flames or in pieces?'[67]

'Grid' Caldwell ridiculed any notion that his senior flight commander was in any way affected by having fought on an almost daily basis for three months: 'I cannot say that during the time I knew Mannock well, as when he was with 74, I sensed any signs of him cracking up. He never "went sick" or asked for time off... He was in good health when he went on leave.'[68] Nevertheless, in a hectoring and brooding letter to Jess Mannock on 16 June, her brother admitted that:

> Things are getting a bit intense just lately and I don't quite know how long my nerves will last out. I am rather old now, as airmen go, for fighting... These times are so horrible that occasionally I feel that life is not worth hanging on to myself, but 'hope springs eternal in the human breast.' I had thoughts of getting married, but...?[69]

In a late 1930s collection of newspaper articles Ira Jones confirmed that by this time Mannock's 'nerves were frayed by the tension of war'.[70]

James Dudgeon created a succession of confessional scenes, but with Jones long since dead, it is hard to see on what basis – none of the latter's three books mentioned either of the incidents described. First, Mannock responded to his second medal in a week by telling Jones that only loyalty to Caldwell was keeping him in the air, and then, on the night of Cairns's death, he averted a complete psychological collapse only by the need to exhort and entertain the rest of the squadron. In scenes worthy of Vernon Smyth at his most creative two decades earlier, Dudgeon portrayed the tortured soul hurtling at breakneck speed to his inevitable destiny.[71] Mannock was clearly under enormous stress and in urgent need of a rest, but there is insufficient documentary evidence to support the notion that by the time he left Clairmarais he was already in the first stages of a nervous breakdown. He may well have been, but the truth is that we simply don't know. What is clear is that by the time Mannock arrived in Wellingborough he was displaying all the signs of what the War Office at the time labelled 'neurasthenia', and what today would be termed 'war trauma'. Jim Eyles was shocked by his friend's general demeanour, not least an apparent inability to control the shaking and twitching in his hands.[72] Mannock was only home for a fortnight, and the first week he had been in London and Birmingham, so the time spent with the Eyles family was actually quite short.

Jim Eyles was convinced his friend 'was in no condition to return to France', especially given that he was not going back to his old squadron. Immediately prior to leaving for England Mannock had learnt from Caldwell that Colonel van Ryneveld wanted him to replace Bishop at 85 Squadron, and that quite frankly he had no choice. Bishop would be departing forthwith, and it was vital to replace him with a squadron commander of similar stature. 85's pilots, having previously vetoed McCudden, were unanimous in their choice of Mannock. Unlike Bishop, he would provide leadership on patrol as well as on the ground. Mannock's posting would be as of 21 June, even though he would not actually assume his command until after returning from leave. Waiting at the RAF Club would be confirmation of the posting, and of his promotion to major.[73] In the club or the mess Mannock could still maintain a facade, but in the kitchen at 183 Mill Road he became quite literally a quivering wreck:

> His face, when he lifted it, was a terrible sight. Saliva and tears were running down his face; he couldn't stop it. His collar and shirt-front were soaked through. He smiled weakly at me [Eyles] when he saw me watching and tried to make light of it; he would not talk about it at

all. I felt helpless not being able to do anything. He was ashamed to let me see him in this condition but could not help it, however hard he tried.[74]

Mannock later tried to make light of the incident, and subsequent conversations studiously avoided any reference to either what he had left behind in France, or what would be waiting for him upon his return. When the time came for him to leave, Eyles sensed that this would be the last occasion on which he would see his friend alive. There was an implicit shared assumption that this really was goodbye, and the usual jokes and promises were noticeable by their absence. When Eyles later recalled how 'that last leave had something very final about it', the tone of his remarks negated any hint of hindsight. Patrick's wife, Dorothy Mannock, present at a farewell party for her brother-in-law, shared a similar sense of foreboding and inevitability.[75]

Mannock's mental condition was certainly consistent with the behaviour of other pilots and observers who experienced either total or partial nervous breakdown, as became clear in evidence given to the 1920–22 War Office Committee of Enquiry on 'Shell-Shock'.[76] The committee was set up to clarify how 'shell-shock' related to the Army's traditional notions of cowardice. The report's conclusion was something of a fudge, reasserting the existence of cowardice, but accepting that sometimes an individual technically guilty of being a coward might in fact have been psychologically damaged and therefore incapable of obeying orders.[77] Drawing upon his celebrated work at Craiglockhart War Hospital, the neurologist W.H.R. Rivers rejected the term 'neurasthenia', preferring to discuss the condition of 'anxiety neurosis'. Conscientious soldiers tried to deal with their anxiety by repressing their deepest fears, at least when in public, but this only rendered the situation worse, sometimes even encouraging hysteria. Mannock's strategy for dealing with his 'flamer' obsession, by publicly engaging in black humour, was consistent with Rivers's hypothesis. Rivers believed that those officers most vulnerable to anxiety neuroses (tears and tremors, insomnia and amnesia) were ex-public school boys conditioned to display self-control and repress fear. Conversely, those soldiers able to acknowledge that being fearless was an impossible ideal were the best qualified to survive. As a soldier's son Mannock faced up to his fears only in private, or publicly through macabre behaviour, so in this respect he had more in common with the upper middle class than with his proletarian peers. Rivers also suggested that within the RAF scout pilots were often the least likely to experience 'psycho-neurosis' because unlike, say, an observer, they were in control of their machines. Such an assertion was consistent with the emerging consensus that neurosis was a functional disorder peculiar to industrial/trench warfare, where passive soldiers felt besieged, immobilized and vulnerable until such time as the war was no longer static.[78] However, when an RAF psychiatrist, Squadron Leader E.W. Craig, listed 18 'causes which, in his judgement, led to nervous breakdown in the flying personnel', nine were directly applicable to Mannock's particular circumstances – including four of the top five. Regular flying at low altitude, for example while ground-strafing, was especially debilitating, but worst of all was: 'Flying at high altitudes for prolonged periods, cold, oxygen want, failure of respiratory mechanism, cardio-vascular and nervous systems leading to breakdown.'[79]

In *A Psychological Retrospect of the Great War* W.H. Maxwell noted that, whenever anger could not be expended against the enemy, it was targeted towards those in authority, most especially the staff. Such antipathy, while for most a convenient means of tension release, could for some have serious psychological consequences. Mannock's scorn, public and private, for 'the brasshatted, red-tabbed "powers that be"', was increasingly inseparable from his obsession with burning to death – by the spring and summer of 1918 most front-line pilots were convinced that only the generals stood between them and the universal adoption of parachutes.[80]

Both J. T. MacCurdy in *War Neuroses* and the Cambridge psychologist F. C. Bartlett in his postwar training lectures for officers argued that a combination of prolonged fatigue and acute stress provoked periodic disorientation *and* a morbid fatalism, whereby death came as a release.[81] In some cases this state of mind spurred the soldier on to acts of great courage. Thus, by 1917 'Mad Jack' Sassoon may well have come into this category:

> As for me, I had more or less made up my mind to die; the idea made things seem easier. In the circumstances there didn't seem anything else to do.[82]

Bartlett suggested that 'there generally has to be some final accident to bring about the collapse.' Mannock, it must be stressed, never did experience 'the collapse', but as we shall see, the death of James McCudden on 9 July hit him very, very hard. Students of 'anxiety neurosis' such as Maxwell, MacCurdy and Bartlett focused in their writings upon the most extreme cases. Mannock clearly did not come into this category – for all his 'bit of nerves', he was still capable of assuming a squadron command, imposing his authority from the outset and destroying eight more enemy aircraft.[83] One can only speculate how much longer he could have carried on before breaking down. Yet intense pressure was placed upon him – 85 was based at St Omer, so all eyes would be on him as he stepped into the shoes of the great Bishop and set about encouraging the Canadian's hybrid collection of pilots to think and act as a team. From the outset he appears to have to found this a highly rewarding experience, and yet a keen expectation of imminent death remained evident to those closest to him. To Bartlett and MacCurdy this weary resignation would have been *the* critical signal, and yet Squadron Leader Craig's 18 causes/signs of nervous breakdown, each drawn from direct observation, offer more compelling evidence that Mick Mannock should have been a priority for Home Establishment. Bartlett of course would have argued that Mannock's reluctance to leave 74 was *ipso facto* evidence that he was at an advanced stage of 'the anxiety state'.

The contrast is with Gwilym Lewis, who in July 1918, once he had been awarded the DSO, could 'see no reason for staying in this dangerous country any longer'.[84] Within days of requesting a transfer he was en route to a transit hospital:

> I knew I could get home any time I liked as I have been feeling a good physical wreck lately, so I got an MO to examine me, and he stopped me flying.[85]

Once away from the front line, relaxing in the kitchen of 183 Mill Road, an emotionally ravaged Mannock could start to objectify his situation – all the evidence suggested that, yes, on a third tour of duty he probably would die. He didn't go back to fulfil his destiny, or as an act of catharsis, or any similar nonsense. Mick Mannock went back to France, partly because of personal pride and a sense of duty, but primarily because he had no choice. The man who had faced up to Sir David Henderson only six months earlier was unlikely to lobby the Air Staff for an alternative posting on the grounds that he couldn't stop his hands shaking. He literally trembled at the thought of going back to France, but he went.

Life and death with 85 Squadron, July 1918

Mannock arrived at St Omer on 5 July, having stayed overnight with 74 Squadron in order to say his goodbyes properly, and give himself more time to recover from flu. By this stage Spanish influenza was endemic, and in London alone well over a thousand people died of flu and pneumonia in the first three weeks of July. However, there is no evidence to suggest that Mannock was another victim – he was not hospitalized, nor even forced to take to his bed for any length of time, and he greeted Gwilym Lewis with an invitation to a dance.[86] Ira Jones suggested that, for all the superficial jollity, Mannock's overnight stay with 74 was a sombre occasion, not least because the squadron's demonstrable affection ultimately reduced him to tears. Caldwell and Mannock agreed to establish a fresh partnership, with both their squadrons operating in tandem; but first 85 had to adhere to the same basic principles of airfighting which had made 74 so successful in such a short space of time.[87]

In many respects Mannock had an easier job than when he joined 74 as most of 85's pilots were hardened veterans. They had chosen him, and he had no need to earn their respect. In fact the boot was on the other foot, and one or two of the squadron's less distinguished members soon found themselves on the way out. The presence of three Americans made 85 an unusually cosmopolitan unit, even in the summer of 1918.[88] Mannock already knew most of Bishop's motley crew and was used to serving in units where the British were in a minority. Indeed it could be said that, given his background, the new squadron commander had more in common with Anzacs, Africans and North Americans than with his own countrymen.

While in no way belittling individual heroism, the emphasis was now on teamwork, with techniques familiar from 74 Squadron deployed to raise morale, foster interdependence and encourage collaboration. As with A flight, on return from patrol everyone was expected to attend detailed debriefings, where even the most experienced pilots gained fresh insight into the tactics of formation fighting and concentrated attack. For the first two days after assuming command, Mannock confined squadron manoeuvres to behind the British trenches. Then, on the evening of 7 July he led the whole squadron across the lines, his flight acting as a decoy while the other two flew at different altitudes some distance away. Mannock insisted that no attack should take place until he fired a red Very light. Over Doulieu the three decoy aircraft attracted a large number of Fokker DVIIs. The Germans were in effect ambushed twice as successive flights dived into the mêlée. A jubilant 85 returned to St Omer, having destroyed five of Germany's top fighter aircraft, with no losses and only minimal damage from enemy fire. Mannock's methods clearly worked, and he himself had added a further two victories to his tally.[89] Over the next fortnight he kept up a rigorous programme of training, focusing especially upon the youngest and least experienced members of the squadron. As with 74 and 40, when in combat Mannock protected the novices, setting them up to claim their first victory. Personal success appeared now to be a secondary consideration, with Mannock seemingly content to orchestrate proceedings and blood the most recent arrivals. New tactics were experimented with, and in some cases quickly discarded. Thus Jones noted abortive attempts to coordinate destruction of a chosen target, with every member of the flight firing simultaneously. The risk of collision, and more especially of being hit by 'friendly fire', ended that particular experiment. The ploy of every pilot targeting his opposite number in an enemy formation proved equally ineffective: most pilots became confused and disorientated, and Mannock was alone in claiming a 'kill'.[90]

On 10 July news reached 85 Squadron that McCudden had crashed at Aix-le-Château the previous evening. Although still alive when rescued from the wreckage, he never regained consciousness and died in a Casualty Clearing Station three hours later. Within 24 hours, the last of the three McCudden fliers had been buried at Waverns Military Cemetery.[91] Jones, who was still in regular contact with Mannock, drew parallels with his friend's response to Dolan's death two months earlier – except that this time the desire for revenge was even more intense. Dogfights became more prolonged, and solitary patrols became more frequent. To the alarm of 85's more experienced pilots, their previously cautious squadron commander now began to take unnecessary risks. The master tactician too frequently ignored his own advice, throwing caution to the wind: he was flying too low for too long, and most dangerous of all, following his victim down to ground level in order to confirm a 'kill'. In a portent of what was to come, on 14 July, after a ferocious dogfight above Merville, Mannock trailed a downed two-seater until finally it crashed. To quote the characteristically verbose Jones, 'This blood-lust had attacked other great air fighters...and sent them to their glorious death. Mannock's turn was not far off.'[92]

Previous biographers have suggested that the frenzied activity and 'hysterical destruction' of Mannock's final fortnight was largely attributable to the 'blood-lust' of an avenging angel. There seems no reason to question his urge to avenge McCudden's death, even though the latter did not actually die in combat.[93] However, the intensity with which 85 Squadron waged war over German lines throughout this period was largely attributable to the Allied counter-attacks initiated by Foch on 18 July. Each Wing increased its patrols, urging squadrons to penetrate ever deeper behind the German front line and entice the *jastas* into fighting. Once this preliminary offensive had commenced, with the BEF making substantial early gains, the RAF provided air cover and ground support for British tanks and infantry securing the Lys Salient. Crucially, instead of digging in to consolidate their gains, field commanders pushed on. Air superiority and improved communications were key factors in this new-found confidence, with the RAF endeavouring to work closely with those forces spearheading the Allied advance.[94] Quite simply, 85 was a front-line squadron whose duty it was to take the battle to the enemy, hence its CO's destruction of four more aircraft during the crucial first phase of the British counter-attack from 19 to 22 July.

On 20 July Mannock attended 74 Squadron's farewell dinner for Gwilym Lewis, and chided George McElroy for flying too low. McElroy had by now secured 47 victories, but Lewis pointed to a superior record in protecting his flight – in seven months not a single pilot had died under his command ('Mac I believe thinks it is rather a bad sign, but I am truly thankful!').[95] Within a fortnight the Irishman would be dead, shot down by anti-aircraft fire just as his old tutor had feared. Yet, with cruel irony, by the time McElroy's machine crashed to earth, Mannock himself had already suffered a similar fate.

Lewis's 'great binge' gave Mannock another opportunity to talk with Jones. A few days later, on the afternoon of 25 July, they met up again for tea. Jones and Clements drove across to the aerodrome at St Omer, and were suitably impressed by 85's *esprit de corps*. Mick was clearly the hero of the hour, and for the first time he looked the part. Photographs of Major Mannock suggest that he was again taking pride in his personal appearance; with his cane, polished boots and Sam Brown, and freshly pressed uniform, he looked every inch the commanding officer. Group photographs of 85 and their guests confirm that the mood was upbeat, the morale high and the weather glorious. Missing is the squadron commander himself, and the two VAD nurses who walked over from the Duchess of Westminster's Hospital to picnic on the grass outside the mess.[96] Vernon Smyth believed that one of these nurses was an Irish ward sister Mannock intended to marry after the war. Sister Flanagan is mentioned on a number of occasions in *Ace With One Eye*, but it is obvious from the paucity of detail that Smyth knew absolutely nothing about her. She was a romantic interest who had survived the transition from film screenplay to 'faction' biography. Her existence was probably based upon a passing remark by Ira Jones when questioned by Smyth in the final years of his life. Despite the absence of any solid evidence to link Mannock with Sister Flanagan, Dudgeon nevertheless went along with the notion that she was one of the two nurses invited to take tea with 85 Squadron.[97]

In the course of the meal Mannock embarrassed Donald Inglis, a short, stocky New Zealander new to the squadron, by asking him if he had 'got a Hun' yet. Mannock would almost certainly have known the answer, and when Inglis duly replied 'no', he was told that this was as good a time as any to make his first 'kill'. Excusing themselves from the picnic, the two men made ready to fly. However, Inglis was unable to follow Mannock up into the air, the elevator wheel on his aircraft being jammed tight. Realizing that something was wrong, Mannock headed east on his own. The tea party carried on, albeit in a less relaxed mood as everyone was aware of how long their host had been gone. Finally, after two hours Mannock was back, his guns still cold. Having reprimanded Inglis's mechanic and insisted that they would cross the lines at first light, he retired for a bath. Jones, worried by his mood and behaviour, followed him.[98]

In *King of Air Fighters* Jones claimed that, in the privacy of his hut, Mannock admitted that he was in danger of buckling under the strain, not least because of a firm conviction that death was imminent. He probably did say something to this effect, but it is hard to believe that his exact words were:

> 'Old lad, if I am killed I shall be in good company. I feel I have done my duty... Don't forget that when you see that tiny spark come out of my S.E. it will kindle a flame which will act as a torch to guide the future air defenders of the Empire along the path of duty.'[99]

At which point, arm in arm with an old and trusted comrade, Mannock made his way to the mess, singing 'Rule, Britannia' at the top of his voice. As will be seen, Mick Mannock's final testament owed a lot more to Jones's political agenda than to any faithful record of a fighting man who knows his time is up. The florid prose and fertile imagination of *King of Air Fighter*'s final chapters inspired later biographers, not least Vernon Smyth and Frederick Oughton, who by this late stage in their book might just as well have been writing a novel. James Dudgeon tried to exercise restraint, but even he couldn't resist the temptation of 'Danny Boy' plaintively greeting the dawn and the soothing song of a blackbird to bid the intrepid pair of aviators good luck.

Where Dudgeon does impress is in the detective work undertaken to establish exactly what did happen over Pacaut Wood at first light on the morning of 26 July – the day a nervous and untried Kiwi lieutenant set out with his squadron commander in search of that first elusive 'kill':

My [Inglis] instructions were to sit on Mick's tail, and that he would waggle his wings if he wanted me closer. I soon found that I didn't have much chance of looking round, as Mick would waggle, and the only thing I could do was to watch his tail and stick tight, as he was flying along the lines at about thirty to fifty feet up and not straight for more than a few seconds, first up on one wing tip, then the other. Suddenly he turned towards home full out and climbing... a quick turn and a dive, and there was Mick shooting up a Hun two-seater.[100]

A Junkers CL1 monoplane was on a reconnaissance patrol above a line of trenches stretching from Mont Bernanchon to Pacaut Wood (between Bethune and Hazebrouck). With both sides in such close proximity, and the front line by no means clear, the Germans regularly reconnoitred the British positions to see if there had been any movement during the night. At around 5.30 a.m. the Junkers pilot and his observer found themselves attacked by not just one SE5a, but two.[101] As the lumbering two-seater turned eastwards, its crew must have known they had no chance of escape. As usual Mannock waited until the target filled his sight, and then he fired a long burst – like McCudden, neutralizing the enemy by first killing the observer. He pulled away, leaving his companion with an easy target. Inglis, in his own words, 'got in a good burst at very close range'. He 'flushed the Hun's petrol tank and just missed ramming his tail as it came up when the Hun's nose dropped.' The flaming Junkers spiralled out of control. Fire and a pall of black smoke signalled the machine had crashed into the ground near the shattered village of Lestrem. A shocked but jubilant Inglis signalled success to Mannock. Totally ignoring the advice he had hammered into his pilots day after day for nearly a year, Mannock made no serious attempt to gain height rapidly, crossing the charred remains of Pacaut Wood at around 40 feet. Immediately above, Inglis circled at about 100 feet. German rifle-fire was now joined by a machine gun, and a burst of tracer hit the port side of Mannock's scout, fulfilling his worst nightmare. The fuel tank ignited, and a tiny flicker of bluish-white flame quickly erupted into a cascade of fire, engulfing the fuselage. Struggling to control the burning machine, Mannock turned right, but before reaching the British front line he lost control and the SE5a veered left. It glided down over Pacaut Wood, and crashed behind the German lines, near Merville. Amid a hail of tracer bullets a horrified Inglis circled above the smoking wreckage before gaining height and turning for home. With his machine heavily shot up, and aviation fuel pouring out of a shattered tank, he came down at St Floris immediately behind the British front line. Dragged from his aircraft screaming and shouting, Inglis was soaked in petrol and near hysterical: 'all I could say when I got into the trench was that the bloody bastards had shot my Major down in flames.'[102]

**Untroubling and untroubled where I lie
The grass below, above the vaulted sky.**

John Clare

At St Omer all of 85 Squadron anxiously awaited Mannock and Inglis's return. They all knew that the two men were long overdue, and at 8 o'clock their fears were exacerbated by a report from an anti-aircraft battery at Hazebrouck that Mannock – familiar from the red streamer on his tail – had been seen in action at around 5.30 a.m., but had then disappeared from view. Half an hour later came confirmation from the company of the Welsh Regiment which had rescued Inglis that Major Mannock was down. There was scarcely any doubt that he was dead, and when Inglis returned to file his combat report it was simply a case of establishing precisely what had taken place.[103] A.C. Randall, the senior flight commander, assumed command. He ordered his adjutant that Patrick Mannock, currently stationed in France with the Tank Corps, be informed at once. The young Lieutenant Cushing fulfilled his difficult task admirably, emphasizing Mannock's popularity and 'patience in teaching others...for he was endowed with the spirit of leadership, and any of his Pilots would cheerfully follow him, no matter where he led'.[104] Writing to Patrick two days later, Padre Keymer similarly stressed his old friend's key role in ensuring an Allied victory in the air. Malcolm McGregor, although shocked by the loss of his CO, was nevertheless shrewd enough to note that, 'unlike other stars, he left behind all the knowledge he had, so it is up to the fellows he taught, to carry on.'[105]

That night representatives from every squadron in and around St Omer joined 85 in seeing off their leader. Caldwell, desperately trying to keep spirits high, was in fact devastated. Jones noted in his diary that trying to be cheerful 'was a difficult business':

> The thought of Mick's charred body not many miles away haunted us and damped our spirits. There was more drinking than usual on these occasions; the Decca worked overtime; we tried to sing, but it was painfully obvious that it was forced, as there was a noticeable discord. We tried to prove that we could take a licking without squealing. It was damned hard.[106]

It was damned hard for Julia Mannock, who three days later received a telegram from the Air Ministry informing her that Major E. Mannock DSO MC Royal Air Force had been reported missing on 26 July. Two months later Edward's name was entered on the War Office's register of death as believed to have been killed in action, but no notification was sent to his mother, by this time again resident in west Belfast.[107] This was despite *The Times* having assumed the 'great airman' was dead as early as 9 August.[108] Julia wrote to the Air Ministry on 23 September, and from Birmingham on 24 April 1919, each time seeking confirmation that her son was dead. In early May a further letter, this time from Nora on behalf of her mother, pleaded for further information. Replies to mother and daughter each expressed sympathy for their loss, stating that 'for official purposes' Mannock was dead. Yet when later in the year Cuthbert Gardner requested, in the absence of a death certificate, 'a certificate of the remission of estate and legacy duty', the Air Ministry and the War Office refused, suggesting that the letter of condolence sent to Julia the previous May would suffice. The Canterbury solicitor's repeated requests for formal confirmation of Mannock's death may have been prompted by the family discovering that Edward senior had inquired whether his son had left a will. The Air Historical Section was insistent that it had no definite knowledge of Major Mannock's death, hence the Air Ministry's caution. However, in early 1921 an RAF intelligence officer quoted a German Death Report that Mannock was buried 'about 300m. northwest of La Pierre au Beure, near the road to Pacaut'. The RAF assumed the German statement, like Inglis's combat report, to be accurate, and took action accordingly. All information available as of January 1921 was forwarded to the Imperial War Graves Commission (IWGC), and the file on Mannock's death was to all intents and purposes closed.[109]

James Dudgeon regarded Private Edward Naulls's eyewitness evidence as more reliable than the inexperienced Inglis's combat report, or the account he sent from New Zealand to Ira Jones in 1934. Naulls claimed to take a keen interest in the aerial battles taking place above his head, and was able to identify the little known Junkers CL1 (which Inglis was unable to do). On the basis of the Essex private's testimony Dudgeon argued that Mannock did not shoot himself once the cockpit became enveloped in smoke and flame – with the aircraft at such low altitude he must still have been wrestling with the controls as otherwise it would have crashed earlier. Also, the fire was concentrated on the port side, but the Colt was easily accessible on the starboard side of the cockpit, had Mannock chosen to use it. Whether, in those final desperate seconds before his machine hit the ground, he did at last put a gun to his head remains unknown. Similarly, despite the Air Ministry closing the file in January 1921, no one knows for sure what happened to Mannock's body. The one clear fact is that it was not consumed in the blaze: when Patrick received from the Red Cross his brother's tunic, notebook, service revolver and ID discs, all the items were charred but not seriously damaged.[110] Thus German soldiers definitely removed the body of the dead pilot and must therefore have buried it. Whenever Jim Eyles questioned the IWGC as to his friend's final resting place, he was informed that the Graves Registration Units had been unable to find a grave at the location indicated by the RAF's intelligence report: '300 metres north-west of La Pierre au Beure on the road to Pacaut'. Because of Mannock's postwar fame and reputation, the Commission returned to the area later in the 1920s and carried out a second search. Enquiries in Berlin produced no fresh information, and the IWGC's Paris representative visited the supposed site of Mannock's interment only to find the whole area had been freshly landscaped – in the process of which no fresh graves were discovered. The Commission did reveal that an unknown pilot's body had been exhumed in the neighbourhood, albeit some distance from where Mannock was ostensibly buried. The Graves Registration Unit had been unable to identify this anonymous aviator, and that meant the Commission had to remain neutral over the question of whether he might in fact be Mannock.[111]

Dudgeon suggested that the original German report had been mistranslated. He pointed out that the location description did not match the actual geography of the area. However, by substituting 'north-east' for 'north-west' it was possible to map accounts of Mannock's death onto the known location of a crashed aircraft, from which the 'unknown British flying officer' had been exhumed for reburial by the IWGC.[112] Another Mannock enthusiast, Andrew Saunders, was similarly convinced that the grave dug by the Germans for this unidentified RAF pilot must have been that of Mick Mannock. If this temporary grave had initially been marked then any identification could easily have been destroyed when the area was subsequently being fought over. Dudgeon and Saunders independently arrived at the same conclusion, that when in 1920 'An Unknown British Airman' was reburied alongside George McElroy in the small military cemetery at Laventie, the IWGC was in fact giving Mick Mannock his final resting place.[113]

Dudgeon's suggestion that the original German report was mistranslated is persuasive, despite his wrongly assuming that it had originated from information passed by the Red Cross to the IWGC. It is not inconceivable that a harassed service interpreter, faced with literally thousands of hastily scribbled reports of enemy deaths, may have made a mistake. Similarly, within the IWGC itself clerical errors often occurred, if only because the task of the Commission in the immediate postwar period was so enormous. Its staff had to deal patiently and sensitively with thousands of families of the deceased, all of them desperate for news. Jim Eyles was a regular correspondent in the 1920s, but the Graves Registration Units could only go by the limited intelligence gained from the Red Cross or the Service Ministries. They had no time to seek alternative locations if their information did not produce results. Around half a million unmarked British graves had to be accounted for, the overwhelming majority of them on the Western Front. Over 200 000 bodies were never recovered – witness a never-ending roll of honour on the Menin Gate or at Thiepval – or on the memorial wall at Arras, where Major Edward Mannock is only one of thousands listed among the missing dead.[114]

6
Conclusion

Remembering Mick

> Yet the dawn with aeroplanes crawling high at Heaven's gate
> Lovely aerial beetles of wonderful scintillate
> Strangest interest, and puffs of purest white –
> Seeking light, dispersing colouring for fancy's delight.[1]

Mystery still surrounds Mick Mannock's death and burial: just how did he die, and does his body lie in the grave of 'An Unknown British Aviator' at Laventie?[2] Other questions remain unanswered, not least whether he truly was the 'ace with one eye', and exactly how many enemy aircraft he shot down. Almost certainly there is no definitive answer to any of these questions.

Nevertheless, in the absence of any hostile aircraft other than the two-seater destroyed by Mannock and Inglis, eyewitness claims that he was hit by tracer fire from a German machine gun emplacement appear incontrovertible. Those admirers insistent that no rival was capable of killing Mick Mannock can draw comfort from the fact that it took enemy ground fire to bring down their man. On the other hand such enthusiasts have to explain why he was flying in such a reckless fashion, ignoring every precept that had previously governed his approach to the war in the air. At least one Mannock aficionado, Angus Stewart, believed him to be so depressed that he deliberately acted in a suicidal fashion. 'Grid' Caldwell, who in any case believed there was nothing wrong with his erstwhile flight commander, ridiculed Stewart's suggestion, insisting that Mannock disregarded his own guidelines solely because he had promised Inglis an easy target: 'that low an attack on a two seater so close to the ground. Quite out of character.'[3] Mannock was by no means the first air ace to fight on for too long and in consequence start taking unnecessary risks. However, that in no way suggests he was deliberately courting death. His psychological condition was not so severe that he consciously set out to end his life, but it did mean that he anticipated the end sooner rather than later. There is a fine line between feeling suicidal, and throwing life-preserving caution to the wind – but that fine line nevertheless does exist. A number of reasons might explain why such an experienced yet exhausted pilot continued to fly so close to the ground (exhilaration? impatience? indifference?), but the desire to die is, at least in this case, not one of them. If Mannock was that keen to end his life, then why did he make such a strenuous effort to cross the British front line? However, when finally it was clear the game was up, did he reach for his revolver? James Dudgeon clearly thought he did, and one can understand why. Mannock had been so insistent a bullet in the brain would pre-empt the final agony that he must surely have seized the opportunity to do so.

Similarly, one can be persuaded by the argument of Dudgeon and other Mannock devotees that the IWGC's Graves Registration Unit unknowingly buried their hero at Laventie in 1920. There was definitely a body, and yet that body was never recovered and identified. Indeed no body was discovered at all, suggesting that the original intelligence may have been flawed. Yet not that far away an anonymous RAF pilot was exhumed, who, given the absence of any means of recognition, could well be Mannock.[4] Without dental charts and DNA profiling the argument that the 'Unknown British Aviator' is in fact Mick Mannock will always rely on circumstantial evidence, but clearly a strong case can be made.

Far less convincing is the notion that the 'ace with one eye' repeatedly tricked medical officers in order to train as a scout pilot. While accepting that in the latter part of World War Two Japan's Saburo Sakai returned to combat after losing an eye, he is believed to be unique – and his fame rested on the 60 victories scored before he was wounded rather than the brief but glorious career that followed his return to operational duty. William MacLanachan was clearly convinced Mannock had lost the sight of his left eye, and Ira Jones, while never as explicit as 'McScotch', by no means rubbished the idea. On the other hand, H.G. Clements flew wingman to his flight commander almost every day for nearly four months, and saw no indication whatsoever that Mannock's vision was impaired.[5] Caldwell was equally sceptical of any notion that his best pilot was blind in one eye, even if it 'did look a bit different'. He did, however, acknowledge that the amoebic infection picked up in India might have left a slight deficiency ('but this certainly did not cramp his style').[6] Every pilot who flew with Mannock, including MacLanachan, testified to his all-round vision and his ability to spot enemy aircraft at great distance. It is hard to reconcile this collective memory of an eagle-eyed stalker with the image of a one-eyed daredevil, perpetually fearful that an unforeseen accident may at a stroke end his flying career. The 'ace with one eye' may be a powerful and attractive myth, but it lacks conviction – how could a scout pilot survive for so long with such a critical loss of depth of vision, let alone the obvious reduction in peripheral vision? Reality may be a lot less glamorous, namely that Mannock had a slight deficiency in his left eye, perhaps a mild optical astigmatism, but it could easily be disguised, not least because he relied upon his right eye for accurate shooting.

The other unresolved question is of course exactly how many 'kills' can be attributed to Britain's most successful fighter pilot. The most enthusiastic students of the war in the air, notably members of the Society of World War I Aero Historians, take this issue very seriously. In this respect they are perpetuating a controversy that dates back to discussion within the Air Ministry in 1919 regarding the posthumous award of the Victoria Cross. Mannock was awarded the VC on 18 July 1919 as a result of intensive lobbying of the Secretary for Air, Winston Churchill. Ira Jones encouraged fellow members of the RAF Club to sign a petition, unaware that Jim Eyles had taken similar action in Wellingborough. In Canterbury, the Mayor and the Dean of the Cathedral both endorsed a petition drafted by Cuthbert Gardner and Ronald McNeil. Gardner had retained close contact with his former protégé right up until his death, and he was executor of Mannock's will. Extracts from their wartime correspondence were included in the *Kentish Gazette*'s report that Canterbury's aviator hero had been awarded the VC.[7] McNeil was a barrister turned journalist, who edited the clubland broadsheet the *St James's Gazette* in the early 1900s while seeking nomination in Scotland as a parliamentary candidate. Implacably opposed to Home Rule, he fought and lost four elections and by-elections before securing the St Augustine Division of Kent in 1911. Thus, one of Mannock's most persistent champions was a diehard Ulster Unionist who, even after winning the Canterbury seat, continued to live in Cushenden, County Antrim.[8]

Jones, Eyles, Gardner, McNeil and all the other campaigners for Mannock's VC were kicking at an open door. In May 1919 Churchill had initiated a debate within the Air Ministry concerning the appropriateness of retrospectively recognizing acts of heroism. He had been approached by a former Cabinet colleague, Reginald McKenna. The

Liberals' last Chancellor had pleaded the case for his dead nephew to be awarded the DFC. The RAF strongly objected to this change of precedent, but reminded the minister that in cases of exceptional 'heroic endeavour' a posthumous VC might be deemed appropriate.[9]

Medals won for gallantry in the air between 1914 and 1918 greatly exceeded those awarded in the Second World War: as an example, only one pilot in Fighter Command received the VC.[10] Instead of being awarded for a particular act of conspicuous bravery, as originally intended, it was agreed that the honour could be awarded to RFC and RNAS pilots for sustained acts of courage and leadership. Thus there was a logic in granting Mannock what had previously been given to the three other leading aces: Bishop, Ball and McCudden. Once the unusual nature of these '"cumulative" cases' had been recognized within the Air Ministry, it prompted an anguished correspondence with the War Office concerning their entries in the published list of Victoria Crosses awarded since 1914.[11] Within the Ministry sceptics questioned whether in fact Mannock was being given the VC solely for his achievements in the air following the award of a second bar to the DSO, rightly pointing out that if this was the case then others had a similar claim. Dissent was silenced by the Under-Secretary of State's draft citation for the *London Gazette*. While Mannock's success in destroying seven German scouts in the short period he served with 85 was duly acknowledged, particular emphasis was placed upon his overall contribution to the RAF. Thus he had proved to be an 'outstanding example of fearless courage, remarkable skill, devotion to duty, and self-sacrifice, which has never been surpassed'.[12]

One individual act of courage acknowledged by a VC was when William Leefe Robinson became the first pilot to shoot a Zeppelin down, over Hertfordshire on the night of 2–3 September 1916. The speed with which a relieved government made this award highlighted the important contribution scout pilots could make in the propaganda battle to maintain morale: the Somme had destroyed any lingering glamour attached to waging war, unless conducted up among the clouds or in the desert – and even then 'Lawrence of Arabia' was a postwar phenomenon. Lawrence it transpired was actually a great admirer of Mannock. Yet however much he endeavoured to mythologize his own heroic endeavours, he never subscribed to the view that the war in the air was an oasis of civilized, chivalrous combat.[13] In a long, corrosive and morale-sapping war of attrition the public could gain some small comfort from an idealized, if in fact banal, notion that day after day brave young knights of the air took to the skies astride their mechanical chargers – the dogfight was no anonymous exchange of fire across no man's land but a noble struggle until death. Nor was this idealized view of aerial combat restricted to naive and ignorant civilians safe in Blighty. Eric Leed identified how the myth of the flier was equally powerful on the front line, reminding battle-weary and often disillusioned soldiers why they originally joined up. Thus, by 'assuming the perspective of the flier, the front-line soldier could gain some psychic distance from the crushing actualities of trench war'. Air fighting – and the perspective of the flier, whether real or imaginary – reinvested 'the actuality of war with its initial purposes', imposed order on chaos, and fed the escapist fantasies of chilled infantrymen shivering in damp dugouts.[14] Nor did the notion of an elite of aerial warriors waging a different war from the poor bloody infantry emerge solely as a consequence of entrenchment and massive indiscriminatory firepower. Michael Paris has argued persuasively that belief in the revolutionary potential of aviation, and that 'the men who conquered the skies belonged to a special breed – that of the ultimate hero', was deeply embedded in popular culture well before 1914.[15] Paris rejected the long-held assumption that, in John Buchan's words, 'half the magic of our Flying Corps was its freedom from advertisement', insisting that the 'cult of the air fighter' was encouraged by Sykes, however much Trenchard and even Rothermere may have objected. Thus Buchan was almost unique among tellers of a good yarn in delaying his tale of the RFC until 1919, while short stories, feature articles and popular verse all hammered home the Prime Minister's message that British aviators were the 'cavalry of the clouds' upholding a 'chivalry of the air' long since lost to those trapped in the trenches far below. Lloyd George's florid tribute coincided with distribution in the autumn of 1917 of the War Office's short propaganda film, *With the Royal Flying Corps (Somewhere in France)*. Not surprisingly, Paris wondered whether it was merely coincidental that the War Office's Cinema Committee, liaising with the RFC, had chosen to film 60 Squadron. Among the group of pilots posing in front of their machines was Bishop, already the recipient of the MC and DSO, and whose VC was announced round about the time of the film's general release.[16]

In announcing Mannock's VC the *London Gazette* noted his final tally of 50 victories, although according to 74's squadron records he had passed that figure over eight weeks before he died.[17] Having accepted that Mannock's combat record did indeed warrant the award of a posthumous VC, senior officers nevertheless insisted that his achievements should not be exaggerated. The figure of 50 was calculated on the assumption that, when recommending a second bar to his DSO, Army HQ in France had mistakenly calculated Mannock's tally up to 16 June

1918 as 48, and not the correct figure of 40.[18] There had been no miscalculation, as Headquarters had simply passed on the figure submitted by van Ryneveld, which had of course come direct from St Omer. If anything 48 was an underestimate as, omitting any shared or unconfirmed 'kills', Mannock had downed 51 aircraft by the end of the second week in June. Using the same strict criteria his final score was therefore 61, with three more aircraft destroyed before he went on leave, and nine after he joined 85 Squadron. However that figure rose to 73 if two shared and ten unverified victories were included.[19] These figures were based on assiduous research, including crosschecking with German files and the individual records of all aircraft flown by Mannock. However, long before James Dudgeon and his fellow enthusiasts began painstakingly compiling an accurate list, Mick's friends were already disputing the figure of 50 quoted in the VC citation. Mannock's 'real' score soon began to creep up, ostensibly in the interest of historical accuracy, but more likely so that he could be portrayed as not just Britain's but the Empire's greatest fighter pilot. Despite their initial insistence on ensuring the official figure was correct, neither the Air Ministry nor the RAF bothered to question Ira Jones's readiness to count unconfirmed 'kills', plus any enemy aircraft his friend had been happy to attribute to others in the interest of flight/squadron morale. Thus the final 'unofficial score', as recorded in the VC Gallery at the RAF Museum, is 73 victories, which Jones roughly calculated in the 1930s but nearly half a century later the assiduous Dudgeon subsequently endorsed.[20] Jones's original estimate was by no means a rough guess, based simply upon a justified belief that the Air Ministry had woefully miscalculated Mannock's final tally: 73 just happened to be one more than the total number of victories attributed to Canada's lone wolf, William Bishop.

Ever since *King of Air Fighters* first appeared in 1934, successive British accounts of the air war have implicitly contrasted the generous Mick, happy to pass on a 'kill' to a tyro pilot like Dolan or McElroy, with the tough single-minded Bishop, insistent on claiming every possible victory, all too often on the flimsiest of corroboration.[21] As if to compound this invidious comparison, 85 was portrayed as an ill-disciplined, hard-drinking collection of wild colonial boys desperately in need of the firm discipline and clear leadership which Major Mannock brought to the squadron.[22] This view of Bishop was wholly at odds with his reputation in Canada where he remains a national hero, having in effect founded the RCAF and then returned to oversee its expansion after 1939. Ironically, Bishop's admirers voice a similar complaint to Mannock's acolytes, namely that he should be credited with 'flaming' balloons and with five unconfirmed victories. Thus a Federal website lists Air Marshal Bishop's unofficial final tally as 80, which is five more than the great French ace René Foncke. For the Canadians at least, Billy Bishop really is the Allies' 'ace of aces', on victories alone outstripping Mick Mannock's claim to be the 'supreme air fighter of all nations'.[23] One can hardly be surprised that both Jones and MacLanachan concluded Mannock was the greatest fighter pilot on the Allied side, and yet what is surprising was the complete absence of Bishop from their list of aces. For the British, Ball was the inspiration, McCudden the pioneer and Mannock the perfectionist: he combined the best qualities of the other two, and his innovative tactics left a lasting legacy for the RAF. For Jones and 'McScotch', Mick Mannock truly was the brightest and the best, hence Churchill's laudable decision to recommend the only award worthy of such achievement.[24]

Having seen an Ulster Unionist lobby enthusiastically for Mannock to receive the VC, there was perhaps a further irony. Previous biographers have claimed that Edward senior insisted upon an invitation to Buckingham Palace in order to collect his son's medals.[25] Patrick Mannock was said to have retrieved all three decorations (VC, MC and DFC), with bars, at some point in the late 1920s. He ostensibly bought them for £5 from the woman his late father had bigamously married after abandoning Julia. However, Patrick's daughter insists that this story is wholly apocryphal, and that her father received Uncle Eddie's decorations direct from the King; the medals were thus always in his possession. In September 1992, however, they were sold for £120 000 at Sotheby's, along with the returned identity tags.[26] Two years later, various artefacts not held by the family were put up for auction, including rare photographs of Pat Mannock, the socialist orator and the expatriate engineer.[27]

In Wellingborough the sale of Mannock's medals revived interest in both his military career and his contribution to the life of the town after he had settled in Eastfield back in the spring of 1911.[28] Not that he had ever been entirely forgotten. Only a year earlier, the Gulf War had prompted a full-page spread in the local paper, reminding readers of 'the debt the RAF owes to a former Wellingborough ace'. Predictably, the old canard was brought out and dusted down, albeit to good effect: 'He didn't have the advantage of supersonic speed or any expensive weaponry. He didn't even have the use of one eye.'[29]

In earlier times Wellingborough's association with Mannock had been proclaimed in a much quieter and more dignified manner. The Lutyens-inspired war memorial, built at the Broad Green road junction in 1924, bears on the back wall the name of Edward Mannock, as does the roll of honour in St Mary the Virgin.[30] Since its formation between the wars the local Air Training Corps has been known as 378 (Mannock) Squadron, its first commanding officer being a master at Wellingborough School commissioned into the RFC. Once the RAF had approved the ATC squadron's name, Jim Eyles donated a small aluminium model of a Nieuport Scout, made by Mannock from part of a German aircraft he had shot down. The intention was that cadets would compete annually for the Mannock Trophy, but before they had an opportunity to do so it was stolen, never to be recovered.[31] Jim and Mabel Eyles vacated their first family home in 1924 but remained in Eastfield all their lives. Until he died unexpectedly in 1959, Jim always endeavoured to keep the Mannock flame alive. Even in death the Eyleses were forever linked with Pat/ Mick, their friendship recorded for posterity on the engraved marble surround of the family grave. Around the same time development south of the town centre demanded new street names, and the town council duly trawled its list of local worthies. Today, the Mannock Road estate is a pale shadow of the postwar municipal initiative the man himself would surely have approved of: most of the houses are shabby and run-down, and those that are not have clearly been sold off.

At noon on 23 June 1988, 77 years after Pat first moved in to 183 Mill Road, a centenary commemorative plaque was unveiled by 'Johnnie' Johnson, the RAF's top-scoring fighter pilot in the Second World War. The RAF and the Labour Party could celebrate a mutual hero, but so too could Wellingborough as it belatedly embraced the heritage industry. The town had begun to rediscover and promote its past. By the time attention was drawn to the sale of his medals, Mannock's posthumous portrait had been rescued from semi-obscurity inside the council chamber.[32] Today, the once half-forgotten hero greets everyone entering or exiting the central library – an appropriate location for an archetypal autodidact. Similarly, he enjoys pride of place in two display cases in the Heritage Centre, a gloomy set of rooms hidden away behind the Hind Hotel. This worthy yet uninspiring exhibition dates from 1987, but could just as easily have been established half a century earlier.[33] Wellingborough is clearly proud of its adopted son, but the town appears to have rediscovered his place in local history rather late in the day. The same could not be said of Canterbury where news of Mannock's VC prompted early, but as it transpired premature, discussion of what might be a fitting memorial.[34]

These days the main function of the war memorial in the Buttermarket is to provide a convenient meeting place for visitors to Canterbury Cathedral. The cross stands adjacent to the Christ Church Gate, at the very point where the temporal life of the city gives way to the spiritual. Given that for much of the year the majority of burger-munching tourists appear to be schoolchildren on a day trip from the Pas de Calais, the almost total absence of spoken English is a sharp reminder of the obverse impact the arrival of the BEF had upon an Ypres or Albert. Back in the early 1920s the choice of a runic cross on a memorial located so close to the focal point of the Anglican Communion appeased those Protestants wary of papist iconography.[35] Surrounding that cross are statues symbolic of the fallen, and – whether by coincidence or design – among those listed beneath the aviator is Edward Mannock. His name also appears on the roll of honour in St Gregory's Church and, as with St Mary the Virgin, his is the only name to have wartime honours appended. Such a unique record was clearly inadequate recognition for Cuthbert Gardner. With G.R. Hews, the owner-editor of the *Kentish Gazette*, he campaigned for several years for a permanent memorial to Canterbury's aviator hero. The delay in bringing this about fuelled a popular myth that even in death Mannock had been forced to overcome local and national prejudice. Writing what he claimed to be the first full newspaper profile, in September 1934, a Major Marsden claimed in the *Sunday Graphic and Sunday News* that, 'the citizens of Canterbury... are quite sore about it, having the opinion that if Mannock had been "of the county" much greater publicity would have afforded him'.[36] The old class tensions were still there, not least that between the Yeomanry and the Territorials. Gardner's struggle with civic apathy appears to have been for a variety of reasons, which may have included a certain snobbishness on the part of the city burghers. However, a more likely cause is that in the 1920s attention focused upon public symbols of national mourning, hence the massive programme of constructing and landscaping national, civic or service memorials, and (largely across the Channel) the IWGC cemeteries. The culmination of this phase of universal bereavement came with the inauguration of Edwin Lutyens's monumental Memorial to the Missing of the Somme, at Thiepval in 1932. Individual local memorials, possibly more in keeping with the popular wartime notion that they might have 'practical rather than simply poetic value', could with the passage of time – and in a tastefully understated and non-triumphalist fashion – celebrate as well as commemorate.[37] The Kent and Canterbury Hospital clearly came within this latter category, and complemented construction of the war memorial in the Buttermarket.[38]

When the Kent and Canterbury's initial building phase was complete, Hews sought to extend its memorialist function. The publication in 1934 of *King of Air Fighters*, plus a letter in the *Kentish Gazette* from a 40 Squadron veteran, encouraged Hews and Gardner to launch a public appeal. They secured enough money to provide the new hospital with the 'Major Edward Mannock memorial bed'. Their success encouraged the *Kentish Gazette* to launch a fresh campaign, coinciding nationally with a growing call for air rearmament, which had a particular resonance in the south-east of England. In a surprisingly short space of time Hews succeeded in persuading the Cathedral authorities that Mannock was worthy of a memorial tablet.[39] Perhaps Hewlett Johnson, the famous 'Red Dean', was sympathetic because of Mannock's socialist convictions, but more likely it was the sheer force of local opinion. After all, the tablet states that its presence in the nave is thanks to 'The citizens of CANTERBURY', a hint perhaps of tension between town and deanery. A brief commemorative service had taken place every year since 1919, on the afternoon of the Sunday nearest to 26 July. However, once the memorial tablet was in place it provided a focal point for the Canterbury branch of the Royal Air Forces Association, members of the family, Cuthbert Gardner, and any other friends and admirers who wished to lay wreaths and pay their respects. Appropriately, given its front line role during the Battle of Britain, RAF Manston was represented, and a photograph of the 1946 'Mannock memorial parade' shows well over 50 veterans and service personnel gathered outside the Cathedral awaiting Dr Johnson's address. In 1956 the tablet was belatedly amended once it had been pointed out that the date given for Mannock's death was in fact eight days premature.[40]

Twelve years later the RAFA and the then Dean clashed over his vetoing the Association's suggestion that a Lightning fly low over the Cathedral to mark the fiftieth anniversary of Mannock's death. For the faithful the cruellest irony was that no one had objected to the RAF similarly honouring von Richthofen at an air show the previous April. Nevertheless, over 50 people gathered to hear the Archdeacon pay tribute to the city's bravest son.[41] By this time Patrick Mannock's daughters had rediscovered Canterbury, having left Kent in the late 1930s when the family moved to Edinburgh. Not that their father had been forgotten, his interwar involvement with the Workers' Educational Association in East Kent echoing Edward's passion for self-improvement. It was Patrick's cricketing prowess as much as his brother's aerial feats which had earned them a commemorative seat at St Lawrence's, the county ground south of the city centre. Uncle Eddie's nieces started attending the annual service, which continued into the next century, followed by a brief ceremony at the hospital.[42] The bemused convalescent occupying the Mannock bed found a long Sunday summer afternoon enlivened by the arrival in his or her ward of a small party made up of civic, service and church dignitaries, and a tiny band of immaculately dressed men and women, most of whom were in their late seventies and beyond. The said patient was then presented with a basket of fruit by the Mayor and solemn-faced representatives of the Canterbury RAFA, all of whom were intent on preserving the dignity of the occasion. The numbers attending the cathedral service grew smaller every year, until the RAF veterans, most of them of Second World War vintage, heeded the Dean's advice that after 80 years perhaps the time had come to pay their final respects.

Of the two towns, Wellingborough has in recent times made more of a corporate effort to keep Mannock's memory alive and an integral part of the civic heritage.[43] Major Mannock VC is literally enshrined in the library as a genuine 'collective symbol', the highly visible portrait maintaining an immediacy and an impact. Even the artefacts in the Heritage Centre are at least a token effort to remind locals and tourists alike that here was a real person who lived in the community, went off first to make his fortune and then to fight, and only in death acquired legendary status. The memorial tablet in Canterbury Cathedral is by comparison anonymous, lost amid the grandeur of far more elaborate martial commemoration. It served its purpose but once a year, and then only for 20 minutes or so. Far more visible is 'Mannock House' a large block of maisonettes at the city end of Military Road, ostensibly on the site of the family's first home in Canterbury. Given its proximity to a large and busy roundabout it is hard to avoid sight of a truly uninspiring example of 1960s municipal housing, and yet few if any of the residents – let alone the thousands of passing motorists – would know from whom the building gets its name. The truth is that the overwhelming majority of today's 'citizens of CANTERBURY' have no idea who 'Major Edward Mannock' was – and why should they? One is tempted to suggest that he was never as well established in the popular consciousness as local dignitaries between the wars liked to think. After all, it does not say much for the general level of interest if it took two decades to correct the date of his death, particularly given the 'extent to which the ebbing and flowing of the Great War are determined by the gravitational pull of the calendar'.[44]

So far the commemoration of Mannock has been seen as for purposes of: civic pride; the maintenance of public morale during and immediately after the war; and the perceived need to recall – and keep recalling – both national and individual sacrifice. The notion of sacrifice, at the heart of wartime and postwar bereavement (a justification and a rationalization for the bereavers of the loss of the bereaved), was both a short-term and then a much longer-term phenomenon. In the short term a sense of purposeful sacrifice was a deeply personal comfort, but beyond that early intimacy it became inseparable from an overtly political agenda. Memorials of 1914–18 were a solemn reminder of the evils of war, but it would be a mistake to assume that they were therefore synonymous with indifference to a credible defence policy. By the mid-1930s a local community's act of remembrance might occasion an unequivocally anti-pacifist message, namely that the nation owed it to those who gave their lives in the Great War always to remain vigilant and prepared.[45] The campaign to commemorate Mick Mannock in Canterbury Cathedral took place between the publication of *King of Air Fighters* in 1934 and *Fighter Pilot* 18 months later. Both Ira Jones and William MacLanachan explicitly used Mannock as an inspiration for politicians and military planners confronted with a renewed German threat. Thus, the master tactician was a role model not just for a rising generation of fighter pilots but for Trenchard, Henderson and Sykes's successors on the Air Staff, and for their political masters.[46]

Mannock's interwar influence on Fighter Command was confirmed by Douglas Bader, who entered Cranwell in 1928. Whether you were flying a Hart or a Hurricane, the mantra of 'Always above; seldom on the same level; never underneath' still applied, let alone the insistence on surprise and on close-range shooting – Bader always insisted on machine guns as he felt cannon encouraged pilots to shoot from too far away.[47] While rejecting Bader's advocacy of the 'Big Wing', 'Johnnie' Johnson agreed with him on the legacy of thinkers like Mannock and McCudden. Not only had they left clear guidelines for engagement with the enemy, which still applied despite the much greater speed attained by modern fighter aircraft, but they had also pioneered formation flying prior to the point of attack. Perhaps Mannock's most lasting lessons were the importance of clear and selfless leadership, and the importance of a technological advantage in order to ensure a tactical advantage. Thus, having acknowledged the enemy's high-altitude superiority, by August 1940 'Our leaders had re-learnt the value of height in the air battle, and even on the few occasions when they possessed an apparent height advantage over the 109s they now rarely failed to leave a covering force of Spitfires high in the sun.'[48] Mannock would of course have taken exactly the same precaution. 'The Few' had a keen sense of history, not least because they had grown up reading of their predecessors' exploits. Literature on the first air war was still a regular source of reference a quarter of a century later. Victor Yeates's 1934 epic, *Winged*

Victory, for example, was highly popular with both fighter and bomber pilots, not least for its resolute refusal to glamorize aerial combat.[49] In 1942 Ronald Adams, controller of fighter operations for 74 Squadron, wrote to Patrick Mannock requesting a photograph of his brother for the mess at RAF Hornchurch. Adams paid tribute to Mannock's unique place in the evolution of Fighter Command:

> All modern fighters hold him in reverence because we regard him as the first of the fighter leaders. There had been, as I remember, the others like Ball, Bishop, etc., but Micky was the first one who really led a squadron into air battle and taught them the tactics on formation.[50]

As in 1918, the 'Tiger Squadron' again boasted some of the most aggressive pilots in Fighter Command, not least the most successful ace of 1940–41, 'Sailor' Malan. 74 had vacated Hornchurch by 1942 when 'Paddy' Finucane arrived as Wing Commander Flying. Finucane, one of only eight Irishmen to fly in the Battle of Britain, warrants mention because the parallels with Mannock are striking: the Anglo-Irish working-class background; the radical (in this case Republican) roots; the Catholicism; the initial disappointments; the careful attention to weaponry and technology; the tactical nous and the consequent success in downing enemy aircraft; the medals and the rapid promotion; the affection and respect generated among those under his command; and the all too brief final posting. The big difference between the two men, other than age (Finucane was only 22 when shot down), was the high profile 'Bader's successor' enjoyed once Fighter Command discovered a flair for publicity and PR.[51]

Back in the mid-1930s the gist of 'McScotch' and Jones's message was that the *Luftwaffe* 'are doing their best to simulate the spirit demonstrated by us during the war', namely a determination always to take the battle to the enemy. Meanwhile a dispirited and unfairly criticized RAF 'shows signs of succumbing to the demoralizing and uninspired system pursued by the Germans during the war'. Von Richthofen had refined the defensive tactics evolved by the *Jastas*' first air ace, Oswald Boelcke, in 1915–16, but this meant surrendering large stretches of the German lines. Enemy pilots were eager to fight, but an inflexible, highly disciplined system denied them the opportunities seized by their British counterparts. Echoing Buchan's Archie Roylance, Jones pointed to 'the cruel, inhuman machine that took everything from a fighter pilot including his individuality. We called it Hunnishness.'[52] *Luftwaffe* pilots in the new Germany, 'now liberated from the cast-iron rigidity of

Prussianist discipline', were being trained by *Jagdstaffeln* veterans such as Udet, but using methods pioneered by Mannock. Well aware that both the Hurricane and the Spitfire were still in the pre-production stage, Jones and MacLanachan gloomily anticipated the *Luftwaffe* being the first to fly a new generation of fighter aircraft – oblique references to the Bf109 constituted a solemn reminder that the German Air Service had only inflicted serious casualties upon the RFC when it enjoyed technological superiority, such as in the spring of 1917.[53]

'The amazing amount of publicity that has been given to the German ace Richthofen and the almost entire absence of any real interest in our own fighters' had not only resulted in a disgraceful indifference to the exploits of Mannock and his comrades, but it had generated a dangerous and debilitating inferiority complex. This 'demoralising and subversive propaganda' was undermining national resolve, and preventing the RAF from upholding the offensive tradition of Ball, Mannock and at a strategic level that of 'Boom' Trenchard. If Stanley Baldwin was right, and 'The bomber will always get through', then the lesson was not to surrender, but to ensure that they were British bombers, complementing the critical role at the home of Fighter Command.[54] *King of Air Fighters* appeared at a moment when, for the first time since 1918, attention was again focused upon a resurgent Germany's potential threat to national security. Thanks to Fleet Street strategists and Wellsian fantasists, mass aerial bombardment had never lost its grasp upon the popular imagination, but now it had once again become a reality. In January 1934 the National Government tacitly accepted Hitler's breaching of the Versailles Treaty's disarmament clauses, proposing to the French that they accept a German air force half the size of their own. Paris rejected the proposal, but as far as Goering was concerned the British were turning a blind eye to rapid expansion of the *Luftwaffe*. That summer Parliament debated whether plans to bring the RAF nearer to its approved strength undermined the credibility of the World Disarmament Conference.[55] Still on active service, Jones insisted that the nation's defences – let alone those of the Empire – had rarely been lower, or the threat greater. From his desk inside the Air Ministry, where he passed his days ostensibly drafting a pamphlet for the Air Historical Section, Jones would have been familiar with the Service Chiefs' alarm at just how exposed British interests were worldwide, let alone much closer to home. A chancer and a wheeler-dealer, whose service career had gone nowhere once he ceased adventuring in the skies over Murmansk and the Middle East, Jones had nothing to lose by blithely ignoring an Air Council requirement that his books and feature articles should be vetted prior to publication.[56] With its last

30 pages devoted to the lessons of the first air war for those presently preparing for the next, *King of Air Fighters* was both a biography and a polemic.

To a lesser degree the same was true of MacLanachan's memoir. 'McScotch' made clear his admiration for Jones, both as a biographer and as a pilot.[57] For both men Mannock was a hero pure and simple, but he was also an inspiration, and an example to British youth:

> To the modern schoolboy, Edward Mannock is the most inspiring figure in the Great War. His achievements may well serve as a beacon to them; a guide along the path of personal endeavour and Duty to their state.[58]

Thus the demands being made of the politicians were twofold: to ensure that the nation's fighting forces – and in particular the RAF – were adequately equipped to face the growing threat across the North Sea, *and* to prepare the next generation of happy warriors for the task ahead. Writing at the same time Cecil Lewis took a far more pessimistic view of the deteriorating international situation. In *Sagittarius Rising* he insisted that aerial defences could do little to prevent systematic destruction of civilian targets, and that it might well take more than one major conflagration for the great powers at last to see sense.[59] Anticipating the later novels of Captain W.E. Johns, the idealistic Lewis insisted that:

> If we really want peace and security, we must pool our resources, disarm, and set up an international air police force, federally controlled. That force must be as incorruptible, free from bias and self-interest, and devoted to law and order as our civil police today. There is no other way.[60]

Lewis and Jones were in agreement that no amount of ARP planning could protect Britain's cities from annihilation, and both were equally out of touch with reality, albeit for very different reasons. Cecil Lewis's internationalism was at the opposite end of the spectrum from the super-patriotic siege mentality of Ira Jones. The latter advocated 10 000 bombers ever ready for the successful pre-emptive strike: 'Such a force will give us the whip hand in European affairs and peace could thus be secure.'[61] Jones did himself no favours by peddling this nonsense to readers of downmarket Sunday newspapers, and yet his efforts to project the aviator as a Homeric hero, quite literally above the toils and tribulations of everyday life, were very much in tune with the spirit of the times.[62]

As Samuel Hynes has observed, Lawrence drew heavily upon the *Iliad* in portraying his Arab tribesmen (and later in the 1920s of course he embarked upon a fresh translation of the *Odyssey*). *The Seven Pillars of Wisdom* depicted heroes fighting a war far distant from the Western Front: the revolt in the desert was a pre-industrial cavalry campaign fought across vast distances, and superficially at least it adhered to a romantic ideal of courage and nobility.[63] Aviation, even after the war was over, enjoyed a similar mystique. Paradoxically, the aircraft was seen as both a weapon of mass destruction *and* as a modernist icon of adventure, liberation, excitement and escape. The fascination with flight of English writers in the early 1930s clearly owed a debt to Yeats. For writers like Auden and Day Lewis the appeal of aviators was that they were classless technocrats: fliers staked their claim to a place in the national consciousness by dint of individual effort and psychological self-discipline, even if the price was repressing their deepest, darkest fears.[64] Whatever their personal Freudian agenda, for once the Auden generation was in tune with mainstream opinion and popular culture. Record-breaking, whether speed or distance, produced a succession of popular action heroes – and heroines, notably Amy Johnson. Flying boats embodied luxury, glamour, romance and a reaffirmation of empire. Even Mr Chamberlain felt a thrill of anticipation each time he commuted from Heston to Germany. Once he re-emerged into the spotlight 'Aircraftsman Shaw' made sure his public world was one of Supermarine seaplanes and super-fast rescue boats. Yet, for all Jones's admiration of Lawrence, he could never envisage him fulfilling the same inspirational role as a Mick Mannock. 'Ross/Shaw's' much-publicized withdrawal from the limelight had rendered him especially attractive to progressives and pacifists. Auden's psychologically tortured 'Airman' in *The Orators* (1932), thrust in to the limelight by a generation and class he loathes and yet is expected to defend, was based on Lawrence. Ransom in *The Ascent of F-6* was similarly assumed to be Lawrence, and Auden happily agreed that this was the case. For writers obsessed by their individual and collective response to the next war, Lawrence was a 'new kind of hero, the agonising, somewhat neurotic intellectual who had transformed himself into a man of action by an act of will'.[65]

Few wartime aviators had any wish, with the return of peace, to reinvent themselves in such an image, and a fledgling and insecure RAF would have been horrified at the thought. Postwar literature, not least the Service's official history, sought to perpetuate aerial combat's spurious association with glamour and romance with considerable success.[66] Even when more balanced accounts of the air war began to appear later in the 1920s, no RFC equivalent of *Memoirs of an Infantry Officer* or *Goodbye To All That* found a niche in the national psyche. If anything, the cinema's return to the skies above the Western Front in the early 1930s served only to underpin and consolidate popular mythology. Thus even *Dawn Patrol*, with its unequivocal anti-war message, portrayed both Allied and German pilots as chivalrous and dutiful 'knights of the air', magnanimous in victory and humble in defeat. Mannock would have hated it, even if, like the doomed hero, he was happy to fraternize with downed Germans – he just didn't want to toast them, as in the film.[67] Hynes saw the popularity of Hollywood movies and pulp fiction as further evidence that, while Lawrence consolidated his status as a cerebral and literary hero, the fliers appealed to a far wider audience – they lodged themselves firmly in the popular imagination, 'and so became the next generation's war-in-the-head, the myth that drew young men...to flying in the next war, and affected the way they thought about their air war, and the way they wrote about it too.'[68]

Antoine de Saint-Exupéry, for example, spent the 1920s coming to terms with having just missed the first air war, and then in the 1930s became increasingly fearful that age would disqualify him from the next one. The novels and travel tales of the interwar years drew upon themes, images, ideas, and metaphors familiar to anyone reading a Cecil Lewis or even a James McCudden.[69] Saint-Exupéry's belief that individual freedom could only be achieved by a keen sense of collective duty and mutual responsibility, and that this was best embodied in the fraternity of aviators, was a humanist's response to Mannock's Christian-based socialism. When he died in July 1944 one young pilot noted that 'our most valued comrade' was a 'fine example of faith...he came to share our dangers in spite of his age... Saint-Exupéry is among the greats, those who face life knowing how to respect themselves as men.'[70] *Flight to Arras*, half combat report and half philosophical discourse, is strangely redolent of Mannock. In denying 'a judgement based on the ugliness of defeat', Saint-Exupéry asked, 'Is a man who is willing to die in flames in the sky to be judged by the blisters on his skin?' War was not about danger or combat, but 'the pure and simple acceptance of death'. Appropriately, at 30 000 feet above Albert, he concluded that, 'A fighter squadron does not kill. It sows death. The seeds grow when it has gone.'[71]

Flight to Arras both inspired and influenced *The Last Enemy*, and a meeting with Saint-Exupéry in New York is said to have reinforced Richard Hillary's determination to fly again, despite the terrible wounds sustained when his burning Spitfire crashed in to the Channel in September 1940. Again, Lawrence proved a role model, exhorting the purgative experience of a return to service anonymity and male camaraderie. In *The Mint* – as yet unpublished, but to which Hillary had access – Lawrence had maintained that fame counted for little in the barrackroom, where a man was judged solely by whether or not he could do the job. Hillary himself insisted that the fighter pilot was an intuitive yet inarticulate being, uncomfortable when lauded by civilians, and eager 'to get back to the Mess, to be among his own kind, with men who act and don't talk, or if they do, talk only of shop'.[72] During the Battle of Britain he had been based at Hornchurch, where 74 Squadron kept the memory of Mannock very much alive. Even after he commenced active service Hillary clung to the notion that being a fighter pilot was 'war as it ought to be ... Back to individual combat, to self-reliance, total responsibility for one's own fate'. Yet, to his credit, even when still at Oxford, he never portrayed air fighting as in any way chivalrous, and the claim that 'it's disinterested' recalled McCudden's efforts to remain detached from the very act of killing.[73] Indeed he insisted that *The Last Enemy* was written precisely because 'the last generation' needed to know that their successors had not fallen for 'the stuff about our Island Fortress and the Knights of the Air': 'we were not that stupid, that we could remember only too well that all this had been seen in the last war but that in spite of that and not because of it, we still thought this one worth fighting.'[74] Having survived Mannock's worst nightmare, albeit at a terrible cost, Hillary joined Saint-Exupéry in 'the pure and simple acceptance of death':

> The fighter pilot's emotions are those of the duellist – cool, precise, impersonal. He is privileged to kill well. For if one must either kill or be killed, as now one must, it should I feel, be done with dignity. Death should be given the setting it deserves; it should never be a pettiness; and for the fighter pilot it never can be.[75]

Hillary acknowledged that the pilots who really understood their machines, and had a genuine aptitude for maths and engineering, were rarely the 'long-haired boys' from varsity and public school. The future lay not with the Oxbridge idealist but with those 'tough, practical men who had come up the hard way', and maintained an easy rapport with their ground crew because all had common points of reference and spoke the same language.[76] These were of course the 1940s generation of Mannocks and McCuddens. Hillary claimed that in 1918 such men – or at least those who had somehow survived – had been granted scant opportunity to challenge the power of the 'old governing class':

> Was there perhaps a new race of Englishman arising out of this war, a race of men bred by the war, a harmonious synthesis of the governing class and the rest of England; that synthesis of disparate backgrounds and upbringings to be seen at its most obvious best in the RAF Squadrons?[77]

Hillary's postwar vision was a deradicalized version of Orwell's 1940 prediction that within 12 months there would emerge 'a specifically *English* Socialist movement': 'Most of its directing brains will come from the new indeterminate class of skilled workers, technical experts, airmen, scientists, architects and journalists, the people who feel at home in the radio and ferro-concrete age.'[78] Of course Evelyn Waugh, writing *Brideshead Revisited* round about the time *The Last Enemy* was published, dreaded just such a future, where *pace* A.J.P. Taylor, 'tough practical men' (and women) were in office *and* in power. Hillary had described himself as 'the last of the long-haired boys', and early in 1943 he too disappeared into the night. Privilege brought with it leadership, and the chance to command a Spitfire flight, or to lead a Guards platoon over the top. Yet an expensive education and a well-cut tunic was, again, no guarantee of survival, as Hillary and the rest of the 'Oxford generation' discovered at great cost – even if in the process they saw themselves as vindicated.[79] A quarter of a century earlier, pilots from a similar background either embraced a new technological egalitarianism or they swiftly perished. In this respect the RFC sowed the seeds of its successor's claim to be a genuine meritocracy. Appropriately, from 1941 Fighter Command was headed by a man who had earned his spurs as a squadron commander on the Western Front, Sholto Douglas. Mannock was the meritocrat *par excellence*, but, having staked his claim to be taken seriously in that first postwar world, sadly he died.

The feeling of resentment explicit in Jones and MacLanachan's interwar biographies was still evident, albeit to a lesser degree, in subsequent books and articles: the sacrifice – the 'supreme sacrifice' – made by Mick Mannock should *never* be forgotten. Frederick Oughton, in his 1966 edition of Mannock's 40 Squadron diary, wrote of 'the most remarkable, and in some ways the most ignored fighter ace of all time', and three years earlier, he had joined Vernon Smyth in complaining of a 'debt' never 'amply repaid'. In a world of missile systems and nuclear weapons Mannock was even more likely to be forgotten, and this grievance simply had to be rectified.[80] Nearly 20 years later, James Dudgeon, a lifetime enthusiast from South Africa long since resident in Edinburgh, set out his stall. The themes were all familiar: loss, sacrifice, inspiration, enigma, inadequate recognition and so on. Dudgeon's claim to authenticity rested on forewords by 'Grid' Caldwell and another veteran of 74 Squadron, 'Clem' Clements, backed up Douglas Bader's enthusiastic endorsement. In 1957 Caldwell, along with Ira Jones, had responded enthusiastically when Vernon Smyth first decided to rescue Mannock from oblivion.

Commander V.F. Smyth RNR was a journeyman actor living in Wimbledon who, in the heyday of British feature films set during the Second World War, judged the time was right to pay cinematic homage to an earlier generation of 'The Few'. Smyth was very keen to ensure every scene and event was authentic, hence his letter to the *Daily Telegraph* in April 1957 requesting advice and information. There was an astonishing response, but Caldwell and Jones were the key contacts. The latter revealed that when *King of Air Fighters* was serialized in the *Sunday Despatch* in 1934–5, 'several film companies were interested'. His agent having not surprisingly turned down a paltry offer of £100 for the screen rights, Jones had approached the then young director, Anthony Asquith. Unlike a decade later, when Asquith shot *Reach For The Stars*, voted by cinema-goers the most popular film of the Second World War, the Air Ministry refused to cooperate. Jones maintained that this was a legacy of Trenchard's insistence upon no special publicity for his pilots, but his own close involvement in the project was almost certainly the real reason why the RAF was so reluctant to become involved.[81]

Throughout 1958–9 Smyth and Jones, who by this time was dying, worked on a detailed synopsis of 'Micky Mannock VC'. Jones's view that Mannock and Lawrence had been drawn from the same mould may well have spurred them on: Terence Rattigan's play for the Haymarket, *Ross – a dramatic portrait*, and David Lean's feature film for Columbia, *Lawrence of Arabia*, had both reached their pre-production stages. Smyth produced a very full outline of a film biography, only one step away from a shooting script. The bulk of the story was a flashback, beginning and ending with 'McScotch' and Jones meeting at Mannock's memorial tablet in Canterbury Cathedral to reminisce about the great man. (Did Smyth have in mind the latter part of Michael Powell and Emeric Pressburger's *A Canterbury Tale* from 15 years earlier?) In an action movie worthy of John Mills and Jack Hawkins, the shadowy Sister Flanagan was to provide the romantic subplot (Virginia McKenna to Richard Todd's Mannock?). Smyth naively sent the manuscript to Twentieth

Century Fox, and then more sensibly approached J. Arthur Rank. When the latter pulled out he saw the writing on the wall, and turned to commercial television. After a brief consideration of Smyth's proposal, Associated-Rediffusion wisely said no. 'Micky Mannock VC' is most definitely not a hidden gem waiting to be dusted down by an enterprising producer.[82]

Some time in the early 1960s a by now desperate Smyth must have come into contact with Frederick Oughton, a freelance writer whose interest in the RFC had resulted in a recent volume of profiles, *The Aces*. Oughton had traced Mannock's 1917 diary to a relative of Jim Eyles in Wellingborough, Mrs Eleanor Woolford, who subsequently donated it to the RAF Museum. Together, Oughton and Smyth turned the lengthy film synopsis into an ostensibly 'honest and objective' picture of the *Ace With One Eye*.[83] What in fact they produced was a dreadful hybrid, drawing upon the worst of popular biography and the worst of popular fiction.

Smyth could gain grim satisfaction from the debt Peter Roberts's radio play *The Action of the Tiger* clearly owed to *Ace With One Eye*. Broadcast appropriately on 11 November 1989 – and again early in 1991, on the eve of the Gulf War – the play's title was clearly an oblique reference to 74 Squadron. Action focused almost entirely upon Mannock and an adulatory Welshman called 'Morgan', who was obviously based upon Ira Jones. Roberts must have read *King of Air Fighters*, but the real inspiration was *Ace With One Eye*. Both book and play share a similar enthusiasm for stilted, unconvincing dialogue. The play's central character bears little resemblance to the real man, being gauche, awkward and guileless. Having achieved success, while still remaining a thoroughly decent chap, Mannock at last confesses to Morgan his innermost fears, and then prepares to meet his fate. The play was awful: the characterization superficial, the dialogue risible and the acting feeble. There was a manifest absence of dramatic tension, and a singular lack of aural ambition in conveying to the listener the reality of life in a front-line squadron in 1917–18.[84] An opportunity was lost to exploit the full potential of radio for feeding the listener's imagination, unlike the 1976 film *Aces High* which, despite obvious budgetary constraints, made a real effort to capture the scale of aerial combat in the final two years of the war. Jack Gold and Howard Baker's demythologizing screenplay, drawing heavily upon *Sagittarius Rising* and *Journey's End*, focuses upon a squadron commander who is clearly not Mick Mannock (the hard-drinking Major Gresham is older than his years, having joined the RFC from public school), but his attitude to flying and fighting demonstrates the same attention to detail, as well as a similarly ruthless view of the enemy. He even shares Mannock's anger at the generals' antipathy towards parachutes. Gresham and his senior officers survive because they know their aircraft (appropriately, the SE5a), always check their weapons, and – even if they don't necessarily trust them – realize the importance of good relations with their ground crew.[85]

While *Aces High* publicly acknowledged the influence of Cecil Lewis and R.C. Sheriff, there was clearly a heavy debt to Derek Robinson's 1971 Booker Prize-nominated black comedy, *Goshawk Squadron*. Major Woolley's emotional condition is verging on the psychotic, and like Gresham he is a very old 23. Given his state of mind, and undisguised sadistic tendencies, Woolley is obviously not Mick Mannock, and yet Robinson had clearly researched his subject thoroughly. Woolley, a stout-swigging, foul-mouthed Brummie, who as a regular might, just might, have made sergeant major, is far more brutal than Mannock in disabusing his younger pilots as to the romance and glamour of the war in the air. Similarly, he is much cruder and more direct in inculcating good habits. Yet Woolley's training methods suggest Robinson was familiar with those pioneered by Mannock at 40 and 74 Squadrons, not least the emphasis on marksmanship as every fighter pilot's paramount skill. Woolley insists every officer loads his own ammunition, and defends the SE5a precisely because it 'is the best gun-platform ever made'.[86] Interestingly, Peter Townsend, who in his account of flying with 85 Squadron noted Mannock's lasting influence, wrote a preface to *Goshawk Squadron* applauding the 'barbarous' Woolley's approach to air fighting.[87]

A decade after *Goshawk Squadron*, and nearly two decades after *Ace With One Eye*, James Dudgeon's slim volume, *'Mick'*, revived interest in Mannock, not least among fellow members of Cross and Cockade International. 'The First World War Aviation Historical Society' was established in 1970 and by 2000 boasted over 1500 members in 25 countries.[88] These trench-touring autodidacts, like their colleagues in organizations such as the Western Front Association, are highly knowledgeable – but rarely able to locate their subject(s) within a wider historical context. The perpetuation of Mannock's memory, ostensibly for noble reasons, clearly cannot be separated from a widespread fascination – largely male – with matters military and in particular flying. Never has the non-professional historian been so well provided for: the proliferation of specialist books, software, CDs and videos; the ceaseless demand for War Office regimental and personnel files at the National Archives; the crowds at air tattoos; and the popularity of specialist societies such as Cross and Cockade. These are all testimony to what critics would contend is an unhealthy masculine obsession with the minutiae of two world wars – an obsession this book sets out to challenge.

Socialist air ace

Even the most macho admirers of Mannock have problems with their man's intense hatred of Germans. Not that this is a recent dilemma – from the outset old comrades laboured to explain away his almost obsessional loathing of the enemy. Even if RFC/RAF pilots were never knights of the air, they were expected to treat their foe with respect and generosity of spirit, and not with utter contempt. Padre Keymer, early confidant and man-of-the-world father-confessor, provided Jones and MacLanachan with what seemed at the time a sufficiently convincing explanation. Keymer, eager to construct an icon, insisted that his friend's 'character was built on a spiritual foundation':

> He was a self-educated man; and like most men of his type, he combined academical knowledge with bitter experience, which resulted in a strong, sound, forceful character... His hatred of the German, which increased with the length of the war, was distinctively more of a hatred bred of civilisation and religion than of the primitive instincts of man.[89]

Jones gratefully quoted Keymer's flimsy attempt to rationalize Mannock's increasingly paranoid view of the Hun. MacLanachan similarly endeavoured to sanitize his behaviour. Thus the early biographers provided a convenient explanation for later books and articles: quite simply, Mannock did not hate Germans, but what they stood for.

Clearly Mannock's prolonged incarceration in Turkey was not enough to explain how someone extolling the virtues of international working-class solidarity could appear to take so much pleasure in the destruction of his adversaries – except on those occasions when investigating a charred airframe or a broken body brought him to his senses. Jones, relying heavily on Jim Eyles's defence of his dead friend's behaviour, implied that it was precisely because he was an international socialist that Mannock 'developed a ferocity and temper unsurpassed by any air fighter in sheer intensity and the deadliness of his hatred for the German.' Victory for the Wilhelmine Reich would, in his mind, have destroyed a 'new and better civilisation, painfully born to life out of the industrial shambles... he saw the whole Liberal-Protestant spirit of Europe menaced by the reactionism of Prussian militarism'.[90] Hence

Mick/Pat went to war because he was a patriot *and* a progressive. Yet it was that same 'Liberal-Protestant spirit', as embodied in the wafer-thin liberal democracy of Edwardian Britain, which, for all his Labourist moderation, Mannock had rejected – and which he still hoped Ireland could escape from, even if the price of Home Rule was blood sacrifice in the trenches (most emphatically *not* in the streets of Dublin). Dudgeon adhered to the Eyles/Jones portrayal of 'that rare being... a man fighting for a Cause': German *Kultur* was anathema to all free-thinking radicals, and its destruction had to be completed before facing up to the bourgeoisie back home.[91] This was a progressive spin upon a similar claim in MacLanachan's *Fighter Pilot*. 'McScotch', who had known several 'decent' Germans at Oxford, claimed to have exercised a moderating influence upon Mannock. If he did then it was only very temporary. Everyone writing about their hero, from old comrades in west Wales to young journalists in East Kent, has clearly felt uncomfortable reconciling a near pathological loathing of the enemy with their image of Mick the gallant and good-natured warrior.[92] A generous conclusion is that

> behind Mannock's frank contempt of the Germans lay a deeper subconscious instinct of distrust and hatred of the aggressive bullying militaristic spirit of Prussianism which was linked in his mature mind to his early struggles... He was fighting for a cause, for a better world order, and he saw the german [*sic*] militarism as the graetest [*sic*] obstacle to that ideal.[93]

William MacLanachan's final assessment, written late in life, does sound plausible – but then so too does the much blunter conclusion of 'Grid' Caldwell that:

> He really hated the Germans – absolutely no chivlary [*sic*] with him; the only good hun was a dead one. I am afraid we rather fostered this bloodthirsty attitude in 74; it helped to keep a war-going atmosphere which is essential for the less tough types.[94]

Thus Mannock's hatred of the Hun can perhaps be explained by identifying a number of factors, none of them exclusive. Firstly, there is, to use his own words, the patriotic socialist's concern for 'The liberty of our Empire and the World', a fear Jones saw as underpinned by deep resentment at past injustices, whether in Turkey or at home.[95] Secondly, there is the Caldwell view that brash celebration of enemy losses boosted squadron morale, and this was always of paramount importance to someone who placed such a great emphasis on *esprit de corps*. Thirdly, there is the suggestion that an uninhibited Mannock was incapable of disguising his emotions, lacking the ultra-professional McCudden's icy detachment, the well-educated MacLanachan and Rhys Davids' empathy with their foe, or the middle-class Ball's reluctance to demonstrate in public either triumph or anger. (This tendency towards uninhibited emotion could of course work both ways – witness his early distress when directly faced with the result of his work. Mannock clearly did not like killing Germans just for the fun of it.) Finally, and most crucial of all, there is the fact that by the spring of 1918 an intensity of mental exhaustion and nervous disorder obviated any degree of sound judgement and emotional detachment on Mannock's behalf. Given this intense degree of 'anxiety neurosis' he was in no fit state to fight on throughout the summer. Eyles rightly knew that, 'He was in no condition to return to France.'[96] A letter sent to Mary Lewis only a few weeks prior to his last leave offers at least a partial insight into Mannock's state of mind:

> I do not believe that war & the *'Great Push'* are things 'rare and superficial'. Don't you see my dear child that strife & bloodshed & physical 'exertion' & mental anguish are all good, gloriously wonderfully beneficial things for the human race, just exactly the same as you and I experience when we are called upon by our sense of righteousness to resist some animal temptation? These boys out here fighting are tempted at every moment of their day to *run away* from the ghastly (externally) Hell created by their Maker, but they resist the temptation (Heaven only knows how strong it is) and die for it, or become fitter & better for it, or because of it.[97]

The sentiments expressed here, while by no means unfamiliar (not least because of their obvious Catholic resonance), do fuel speculation as to what ideas the writer would have brought home had he in fact survived another four months. What a postwar Mick Mannock might have achieved is, not surprisingly, a constant theme of all books and articles about the man. Clearly he left an enduring legacy as an innovative tactician, but any suggestion that he could have built a political career upon his reputation as a war hero is pure conjecture. It is less likely Mannock would have received the VC had he remained alive, but one can confidently assume that, given his war record, he would have escaped the social disorientation so many 'temporary gentlemen' experienced following demobilization.[98] Dudgeon raised the question of whether the Wellingborough ILP's most famous member could have become a 'first-class Socialist Prime Minister, one who might have changed the course of Socialism in Britain'. Wisely he chose not to answer it.[99] Writing in 1934, when Labour's fortunes under George Lansbury were at an unusually low ebb, and drawing heavily upon Jim Eyles's view of his old comrade as the great lost leader, Ira Jones insisted that Mannock would have become a key figure in the PLP (as MP for Wellingborough perhaps?): 'his forensic powers and great personality would have made him a force to be reckoned with, for he would have effectively combined vigorous and penetrating criticism with popularity.' It is difficult to imagine that 'popular Labour leaders and he would have assuredly linked arms'.[100] The gerontocracy of first generation leaders which controlled the PLP until after the 1931 general election would surely have treated Mannock with as much suspicion as they did Sir Oswald Mosley.

Mannock had the proletarian credentials which Mosley lacked, so one can imagine him being more comfortable within the ILP at Westminster, even if he had fought a very different war from James Maxton, John Wheatley and the other 'Red Clydesiders'. He might thus have been an effective thorn in the Labour leadership's side after 1924 as the ILP launched first its 'living wage' campaign, then in April 1926 the demand for 'Socialism In Our Time', and finally in June 1928 the (A.J.) Cook–Maxton manifesto.[101] The fact is we simply don't know, and Ira Jones was probably much closer to the mark when he confessed in 1957 that the Mick Mannock he knew 'was no MP – had not even been near a House of Parliament – he might have gone if he'd lived in Guy Fawkes day!'[102]

However, the comparison with Mosley, if not Guy Fawkes, prompts speculation that Mannock would have found confinement to the awkward squad of the ILP an uncongenial and unappetizing coda to his wartime adventures. Committee room wrangles with Philip Snowden or Jimmy Thomas would hardly have been on a par with lecturing 'Swazi' Howe or 'Nigger' Horn on the finer points of the Immelman Turn. Does that final letter to Mary Lewis, suggest that, if not necessarily emulating Mosley (who was himself briefly in the RFC), Mannock might have been seduced by fascism's fusion of youth, action, dynamism and, above all, the 'spirit of the trenches'? If so, would Mannock, like Lawrence, have merely flirted with the British Union of Fascists, or followed Henry Williamson's example and wholeheartedly embraced the movement?[103] Certainly, Williamson's friend and protégé Victor Yeates articulated the keen sense of betrayal felt by the RAF's most embittered survivors, albeit lacing his cynicism with a heavy dose of anti-semitism: 'This war, declaring itself godly, righteous, a crusade, was tainted and suspect, and the disgusting ignoble civilisation that supported it deserved eclipse.'[104] Again one can only speculate, but it is hard to imagine Mannock betraying friends and comrades for a black shirt and a platform seat at Olympia. It is even harder to envisage him adopting the same 'former comrades-in-arms' attitude towards the Germans after 1933 as did Williamson, Wyndham Lewis and other ex-service admirers of National Socialism. His absolute loathing of the Hun makes it more likely that Mannock would have lined up alongside MacLanachan, Jones and the legion of Great War fliers lobbying inside and outside Whitehall for the accelerated expansion and re-equipment of both Fighter and Bomber Commands.

All this presumes of course that Mannock would not have moved further to the left. However, if the idea of him joining the BUF is implausible, then the notion that he would have nailed his colours to a Bolshevik mast is even more far-fetched. Not that members of the RFC/RAF didn't join the Communist Party – witness Tom Wintringham, who had delayed entry to Balliol in order to serve in the ranks.[105] Fighting with the International Brigade in Spain 20 years later, Wintringham cast an expert eye over the volunteer squadrons defending the Republic. A good number of Republican pilots were mercenaries, but in Valencia in early 1937 there were enough British and American volunteers to form a squadron. Is it too fanciful to imagine Major Mannock taking this motley crew of leftist Anglo-Saxon aviators and turning them into a combat unit as famous as Malraux's *Escadrille España*, and as ruthless as the top Soviet squadrons sent to defend Madrid?[106]

Despite their very different backgrounds, Hugh Dalton is the Labour politician with whom by the mid-1930s Mannock might have had most in common. Dalton's experience as a 'gunner' with the Royal Artillery in France and Italy left him with a lifelong distrust and dislike of Germans, hence his belief that Labour should adopt a less equivocal line on rearmament. The PLP's tacit support of the military estimates after 1937, and its leaders' strident criticism of Chamberlain between Munich and the outbreak of war, enabled Labour to reposition itself as a party of patriotism (even though it continued to oppose peacetime conscription). Consolidation of the left's claim to 'Speak for England' came of course with Labour's acceptance of office in May 1940, not least Ernest Bevin's towering presence on the Home Front.[107] The election victory five years later 'marked a triumph of social patriotism, of improving the condition of the people as patriotic endeavour, over traditional patriotism'.[108] As the Wellingborough Labour Party celebrated in July 1945 Jim and Mabel Eyles must surely have spared a few moments to reflect on the pleasure their old friend would have experienced savouring such a triumph – and such a vindication. Orwell was more sceptical, arguing that for much of the interwar period British socialists had lost touch with the radical patriotism that Mannock had drawn upon in the years leading up to 1914: for too long the left had been scarred by the support a majority of the Labour movement had given to a war which entailed huge sacrifice for precious little reward.[109] Yet the impression that MacDonald and Henderson's successors were unacquainted with matters military, even if they had supported the war effort, was misleading. For all its flirting with pacifism up to 1935–6, prominent among Labour's second generation of leaders were former servicemen who had fought throughout the Great War, often in appalling conditions and in various theatres of operation. Like Dalton, Clement Attlee had a 'good war', but not as the term is commonly understood, with its peculiarly British 1939–45 connotation of being 'a bit of a lark' – adventure, action and escape from the tedium of civilian routine.[110] Attlee had a 'good war' in so far as he served with distinction, was wounded but returned to the line, showed courage when required but never acted in a foolhardy manner, led from the front and by example, and, above all, always put the interests of his men first. In short he was a first-class company commander. The Labour leader – he succeeded the pacifist George Lansbury in 1935 – was invariably referred to by his military rank, and indeed the British Battalion of the International Brigade boasted a 'Major Attlee Company'. Arguably Attlee is open to Orwell's charge of supporting arms for Spain while still questioning the need for rearmament at home. For too long both leader and NEC sought to reconcile the irreconcilable in the interest of party unity – witness the delay until 1939 of a credible policy document, *Labour and Defence*, with its belated endorsement of meritocratic if not always democratic armed forces.[111] In his memoirs Attlee insisted that he had always rejected as risible the notion that 'an inefficient army was less wicked than an efficient one', and that on the basis of personal experience he had made the improvement of service pay and conditions a priority.[112] The First World War had demonstrated that socialists could be as adept as anyone else at inflicting violence on their fellow human beings. In an era of pacifism and disarmament, when many who had opposed Britain going to war in 1914 remained prominent within the Labour movement, this was an uncomfortable reality of life, which could be conveniently forgotten until the Spanish Republic issued a global call to arms in the autumn of 1936.

Unease, even shame, over what took place behind the lines in Republican Spain means that, when looking back across the past 100 years, the left in Britain only feels comfortable recording and applauding military involvement in the 'people's war' – with all the mitigating circumstances of that highly contentious yet highly convenient concept, the 'just war'. From South Africa at the start of the century right through to Kosovo at the end, the Conservative Party and press have consistently exploited Labour's divisions over – and obvious unease with – the notion and the reality of waging war. Explicitly or implicitly Labour's harshest critics have portrayed the party as pacifistic, appeasing and, worst of all, unpatriotic. Michael Foot's fruitless endeavours to resurrect the spirit of *Guilty Men* in April 1982 did little to dent Tory triumphalism, and Neil Kinnock's endeavours to appear statesmanlike in 1990–1 scarcely countered Central Office insistence that here was a man who would never press the nuclear button. An inability in the spring of 1999 to tar Tony Blair with the same brush created a bizarre situation whereby the *Daily Telegraph* was in an unholy alliance with dissident Labour backbenchers in claiming NATO intervention in the Balkans would fail, while at the same time berating the government for harbouring anti-militarists in its serried ranks. Despite the relative youth and almost wholly non-military background of its front bench, the Conservative Party still encourages a false and obsolescent assumption that, by ostensibly sharing the same values and social background as a presumed homogeneous officer corps, it has a unique understanding of service culture and tradition. Yet in recent years only one prominent politician has had direct acquaintance with service life, and he is a Liberal Democrat – Paddy Ashdown. The postwar National Service generation has now followed the 1939–45 veterans to the backbenches and the Lords. Denis Healey and Jim Callaghan demonstrated that six years in uniform was as much of a formative experience for Labour leaders as for their patrician pre-Thatcherite 'one-nation' Tory counterparts. Major Healey's credibility as a cost-cutting Secretary of State for Defence between 1964 and 1970 rested as much on his wartime record in the Mediterranean as on his reputation as an intellectual bully. Tony Crosland similarly drew on five years of fighting to argue against unilateralism in 1960, and 16 years later for a delay in upgrading the independent nuclear deterrent – under Gaitskell Labour had to be electable, but under Callaghan the priority in a shrinking defence budget had to be decent front-line equipment.[113]

But if the Second World War should be acceptable to the left as a formative influence, why not also the First? Mannock's capacity to wage war uncompromisingly, yet remain a credible figure of the left, is a sharp reminder that – Herbert Morrison and Ernest Bevin notwithstanding – the Labour leadership in the mid-twentieth century had been moulded by its experience on the Western Front and in the Mediterranean.[114] When Attlee detailed political ineptitude and military incompetence in the Norway debate on 7 May 1940 he spoke with the authority of a survivor of Suvla Bay and Passchendaele.[115] Correlli Barnett's suggestion that the Cabinet of July 1945 was largely made up of woolly-minded do-gooders scarcely acquainted with 'the real world' is both inaccurate and absurd.[116] Hugh Dalton may never have been a captain of industry (although Stafford Cripps certainly had) but he had captained an artillery battery at Caporetto and on the Somme – varsity and the university of life were not necessarily incompatible.[117]

Mick Mannock, for all his Jude-like faith in learning, never made the varsity, but he could safely claim to have graduated from the university of life. He was intellectually – if not necessarily emotionally – equipped for a postwar career in politics, and for all his loyalty to the ILP it is hard to imagine him being content with a marginal role. The mainstream meant a closer proximity to the corridors of power, and therefore the possibility of ministerial appointment. Once in office there was the chance to initiate ameliorative action, no matter how modest. One can construct a glittering career for Mannock, but it will always be a deceit, rooted almost entirely in supposition. Ben Pimlott once claimed that a biography is a 'deliberate deceit' in that the evidence is always fragmentary, and yet it purports to reveal a total portrait of its chosen subject. This biographical essay makes no such claim, although Pimlott's alternative depiction of biography – as an 'unpredictable picaresque adventure' – does strike a chord.[118] The search for Mick Mannock all too often proved a frustrating experience, not least because the evidence was so fragmentary. The difficulty lay, not in identifying the high-flying warrior in the vanguard of an 'emotional shift to a new technological paradigm', but in singling out his civilian *alter ego* from the semi-anonymous masses of the Edwardian working class.[119] The temptation is to sigh and say 'Yes, Mannock clearly was a socialist, but away from the mess – in the committee room and the debating chamber – what did he *really* believe in?' In the past neither Mick nor Pat (nor Eddie nor Jerry) have freely offered up their personalities to the close scrutiny of either the old comrade, the hack writer or the investigative enthusiast. Yet surrounding Mannock is a magic, a mystique and above all an exceptionally resilient myth, which the professional historian should neither ignore nor look down upon. Nor indeed should the Labour Party, whether old or new. It is easy to sneer at the sparsely attended annual memorial service, but that would be to ignore the genuine solemnity of the occasion, and also its remarkable continuity – like the most resilient acts of remembrance it survived into our century, providing even the youngest present with a direct line back to 1914–18. There was a quiet dignity surrounding the July gathering in Canterbury Cathedral, which warrants revival. However, to salute *all* of Mannock's achievements – in the sky and on the ground, in the mess and on the platform, in khaki and in mufti – the next time the wreaths are laid it would be fitting if, among the Flanders poppies, there lay a single, rather garish and thus undeniably defiant red rose.

Notes

Chapter 1 Introduction

1. Whether focused on or behind the front line, year on year the Western Front continues to generate an enormous body of literature, from the meticulously researched monograph of the specialist academic to the glossy potted history of the journeyman writer. Military history, unlike just about every other branch of the discipline, boasts expertise in individuals who may be professional historians but, by occupation, are not professional academics: re the British Army during the First World War, e.g. the Imperial War Museum's Peter Simpkins, and best-selling freelance writers Lyn Macdonald and Martin Middlebrook. Anyway, at what point does military history *per se* become a cognate area in its own right: 'war studies' or 'defence studies', or even 'peace studies'? The intellectual parameters within which military history traditionally operated have long since been broken down, e.g. Joanna Bourke's insistence that the study of men and combat is a legitimate field of study for the cultural historian: Joanna Bourke, *Dismembering the Male: Men's Bodies, Britain and the Great War* (London: Reaktion Books, 1996), and *An Intimate History of Killing: Face-to-face Killing in Twentieth-century Warfare* (London: Granta, 1999).
2. For example, as well as Middlebrook and Macdonald, John Keegan in his pioneering *The Face of Battle: A Study of Agincourt, Waterloo and the Somme* (London: Jonathan Cape, 1976; Pimlico, 1991), and Richard Holmes in his television series *Western Front* (BBC2, 1999). For a historiographical survey of literature focusing upon the direct experience of soldiers in and out of battle, see Peter Simpkins, 'Everyman at War: Recent Interpretations of the Front Line Experience' in Brian Bond (ed.), *The First World War and British Military History* (Oxford: Clarendon Press, 1991) pp. 287–313.
3. Pat Barker, *Regeneration* (London: Viking, 1991), *The Eye In The Door* (London: Viking, 1993) and *The Ghost Road* (London: Viking, 1995); Sebastian Faulks, *Birdsong* (London: Hutchinson, 1993).
4. Nigel Steel and Peter Hart, *Tumult In The Clouds: The British Experience of the War in the Air, 1914–1918* (London: Hodder & Stoughton, 1997).
5. Cecil Lewis, *Sagittarius Rising* (London: Peter Davies, 1936; Warner, 1994), pp. 16, 44, 228, 331. SE was the Royal Aircraft Factory at Farnborough's originally

pre-war acronym for 'Scouting Experimental'. By 1917 a 'scout' was a fast single-seater fighter aircraft, with an aggressive combat role very different from its predecessors earlier in the war. They were intended merely to 'scout out' developments on the ground and then quickly report back.
6. Joni Mitchell, 'Amelia', *Hejira* (1976)
7. Ira Jones, *King of Air Fighters: The Biography of Major 'Mick' Mannock, VC, DSO, MC* (London: Ivor Nicholson & Watson, 1934; Greenhill Books, 1989); 'McScotch', *Fighter Pilot* (London: Routledge, 1936; Bath: New Portway, 1972); Frederick Oughton and Commander Vernon Smyth, *Ace With One Eye: The Life and Combats of Major Edward Mannock* (London: Frederick Muller, 1963); James M. Dudgeon, *'Mick' The Story of Major Edward Mannock V.C., D.S.O., M.C., R.F.C., R.A.F.* (London: Robert Hale, 1981; paperback edn 1993). 'McScotch' was Mannock's nickname for William MacLanachan, with whom he flew in 40 Squadron from April 1917 to January 1918; Ira 'Taffy' Jones flew with Mannock in 74 Squadron from February to June 1918.
8. Because of the skills necessary to fly and maintain their aircraft, the 291 175 officers and other ranks who served in the RFC/RAF between 1914 and 1918 all needed to have had a sound basic education. A disproportionate number of public school boys sought to become pilot officers until the War Office took action in early 1917 to redress the balance between the RFC and other regiments and corps. John H. Morrow, Jr, 'Knights of the Sky: The Rise of Military Education', in Frans Coetzee and Marilyn Shevin-Coetzee (eds), *Authority, Identity and the Social History of the Great War* (Oxford: Berg, 1995) p. 316.
9. Frederick Oughton (ed.), *The Personal Diary of 'Mick' Mannock V.C., D.S.O. (2 bars), M.C. (1 bar)* (London: Neville Spearman, 1966).
10. The official history records that, 'Edward Mannock was a great formation leader who had the gift of inspiring those who flew with him. He had a keen, analytical mind, and pilots who served under him have testified that he was always thinking out schemes for the tactical handling of a fighting formation.' H.A. Jones, *The War in the Air Being the Story of the Part Played in the Great War by the Royal Air Force Volume V History of the Great War Based on Official Documents by Direction of the Historical Section of the Committee of Imperial Defence* (Oxford: Oxford University Press, 1937) p. 433.
11. Sadly, any papers – including perhaps his prewar diaries – held by the Eyles family, with whom Mannock lodged in Wellingborough, have also disappeared.
12. See the discussion of myth and war, with reference to Roland Barthes, in Angus Calder, *The Myth of the Blitz* (London: Pimlico, 1991) pp. 2–3.
13. Richard Hillary, *The Last Enemy* (London: Macmillan, 1942; Pimlico, 1997).
14. See Alex King, *Memorials of The Great War in Britain: The Symbolism and Politics of Remembrance* (Oxford: Berg, 1998), which includes a brief summation of differing academic perspectives on the purpose of commemoration, pp. 6–7; Adrian Gregory, *The Silence of Memory: Armistice Day 1919–1946* (Oxford: Berg, 1994); Geoff Dyer, *The Missing of the Somme* (London: Hamish Hamilton, 1994; Penguin, 1995); Jay Winter, *Sites of Memory Sites of Mourning: The Great War in European Cultural History* (Cambridge: Cambridge University Press, 1995).
15. Ibid., p. 98.

16. Dyer, *The Missing of the Somme*, pp. 10–11; Winter, *Sites of Memory Sites of Mourning*, p. 98; J.M. Winter in Coetzee and Shevin-Coetzee (eds), *Authority, Identity and the Social History of the Great War*, p. 348.
17. A colleague, when informed that Mannock applied socialist principles to aerial combat, assumed he must have been a Fabian and thus always flew slowly into battle.
18. The War Office's December 1916 assumption of an average active service life for single-seater scout pilots of ten weeks was quickly shown to be over-optimistic. Experienced pilots who survived regular combat served around six months in the front line before leave and a fresh posting, often to a training establishment at home. Ralph Barker, *The Royal Flying Corps in France: From Bloody April 1917 to Final Victory* (London: Constable, 1995), pp. 47–8 and 6; Steel and Hart, *Tumult In The Clouds*, p. 318.
19. Aircraft production more than doubled year on year, 1916–18. It had to in order to meet the demand: to give some idea of the wastage in aircraft at critical points in 1918, of the 1232 aircraft in the air on 21 March, the day Ludendorff launched his great offensive in the west, only 200 were still airworthy three weeks later. On 11 November 1918 the RAF maintained at home and abroad no less than 22 647 aeroplanes and seaplanes. Barker, *The Royal Flying Corps in France*, pp. 251 and 240; Steel and Hart, *Tumult In The Clouds*, p. 346.
20. An effective interrupter gear was first developed by Anthony Focker in 1915, giving the German Air Service an immense tactical advantage until the British and French were able to introduce a similar system of firing forward.
21. Ray Sturtivant and Gordon Page, *The S.E.5 File* (London: Air-Britain, 1996), p. 6. Cecil Lewis preferred the Camel, but his comparisons are with the clearly inferior SE5. Tom Cundall, the thinly fictionalized version of ex-RFC novelist Victor Yeates, relishes the unpredictability of the Camel but balances this against the speed and firepower of the SE5a. Lewis, *Sagittarius Rising*, p. 200; V.M. Yeates, *Winged Victory* (London: Jonathan Cape, 1934; Southampton: Ashford, Buchan & Enright, 1985), pp. 29–30, 44–5 and 53.
22. My thanks to Nigel Steel for his advice on archival material at the IWM, and discussion of his book's contents and ideas prior to publication.
23. '...history is not merely what happened: it is what happened in the context of what might have happened.' Hugh Trevor-Roper, 'History and Imagination', in Valerie Pearl, Blair Warden and Hugh Lloyd-Jones (eds.), *History and Imagination: Essays in Honour of H.R. Trevor-Roper* (London: Duckworth, 1981), p. 364.
24. John Banville, *The Untouchable* (Picador, 1997; paperback edn 1998), pp. 200–1.
25. If one excludes interwar colonial policing, the 1999 war in Kosovo was the first occasion when this was the case, given that the Korean War and Malayan Emergency commenced when Major Attlee was Prime Minister, and the 1960s conflicts in Borneo and Aden were the direct responsibility of Major Healey at the newly unified Ministry of Defence.
26. A label regularly attached to MacDonald by *The March Of Time* newsreel commentators. Adrian Smith, 'Ramsay MacDonald: Aviator and Actionman', *The Historian*, 28 (1990), pp. 14–15; Harold Penrose, *British Aviation: Widening Horizons 1930–1934* (London: HMSO, 1979).

27. If there is an explanation for MacDonald's great love of flying it is the strength of his friendship with aviation enthusiast Brigadier-General Christopher Thomson, ennobled and appointed Labour's first Air Minister in 1924. His second period of office ended tragically on 5 October 1930 when the R101 airship crashed at Beauvais on its inaugural flight. Smith, 'Ramsay MacDonald: Aviator and Actionman', pp. 14–15.

Chapter 2 A Prewar Education, 1888–1914

1. George D. Kelleher, *The Gunpowder at Ballincollig* (Inniscarra, Co. Cork: Kelleher, 1997) pp. 61–72. The barracks were rebuilt at the onset of the wartime Emergency.
2. St Mary and St John's English architect was the prolific and highly regarded George Goldie, his fees – as well as the land on which the church was built – provided by landowners and powder mill directors eager to damp down local Fenianism. Rang IV, Scoil Mhaire, 'The Church of St Mary and St John', *Times Past*, VI (1989–90) pp. 19–23.
3. Jones claimed the newly-married Corringham spent ten months fighting with the Heavy Camel Corps in the 1881–2 campaign to crush Arabi Pasha's uprising against Anglo-French hegemony in Egypt. If he did then it would have to have been *before* he met Julia Sullivan, and not after as stated by Jones (and Dudgeon). Jones, *King of Air Fighters*, p. 4.
4. Records of St Mary and St John RC Church, Ballincollig. In previous biographies Oughton and Smyth listed the wrong regiment, and Dudgeon relied on Jones's incorrect dating of the wedding and statement that Corringhame [*sic*] used his mother's maiden name (which was in fact Riley).
5. Family information; billiards reports in *The Times*, February 1892, January–February 1893, December 1895, March–April 1897, and October 1898. If Mannock bet on himself then he would have lost a *great* deal of money. His name was synonymous with billiards, being used to market cues and billiards tables.
6. 'Employees at Ballincollig Gunpowder Mills', 1815, Ballincollig Gunpowder Mills Heritage Centre; Nora Lynch [Ballincollig local historian] to Adrian Smith, 10 May 1998. Mick Sullivan and Hanora Barrett worked on the Colthurst family's estate at Arbrum, Inniscarra.
7. Major Ralph Legge-Pomeroy, *The Story of a Regiment of Horse: being the Regimental History from 1865 to 1922 of the 5th Princess Charlotte of Wales' Dragoon Guards Volume 1* (William Blackwood and Sons, 1924) p. 190. E.M. Spiers, 'Army Organisation and Society in the Nineteenth Century' in Thomas Bartlett and Keith Jeffery (eds), *A Military History of Ireland* (Cambridge: Cambridge University Press, 1996) p. 340.
8. Records of St Mary and St John RC Church, Ballincollig. The discovery that Corringham was a convert ends speculation that he might have been descended from the Mannocks of Giffords Hall, well-known Suffolk recusants obliged to sell the family seat after the last [Jesuit] baronet was killed in 1798. Mark Bence-Jones, *The Catholic Families* (London: Constable, 1992) p. 62.
9. Edward Spiers, 'The Late Victorian Army' in David Chandler and Ian Beckett (eds), *The Oxford Illustrated History of the British Army* (Oxford: Oxford University Press, 1994) pp. 192–3.

10. Legge-Pomeroy, *The Story of a Regiment of Horse*, p. 193. Unlike the infantry, even nominally Irish cavalry regiments were overwhelmingly English (and Protestant) in composition, hence their value as military aid to the civil power. The total military presence in Ireland was surprisingly small, at around 25 000 men, and the presence of a barracks was often seen by the local community as a welcome boost to local trade – as in Ballincollig. Spiers, 'Army Organisation and Society in the Nineteenth Century', pp. 341–2.
11. Ira Jones stated Edward was born in Brighton on 24 May 1887, and this is the date usually accepted. However, the 2nd Dragoons' regimental history records them as not moving to Sussex until 1888, the year Mannock himself wrote down in his service record. Ever the enigma, he sowed confusion by on other occasions recording his birthday as 24 May 1887 – was the choice of Empire Day *and* Queen Victoria's Golden Jubilee deliberate, intentionally ironic or simply coincidental? He gave yet another date on his passport, and when serving with Mannock in 1918 Jones believed him to be 33. Jones, *King of Air Fighters*, p. 4; Edward Almack, *The History of the Second Dragoons 'Royal Scots Greys'* (London: private publication/public subscription, 1908) p. 84; 'E Mannock', Form M.T. 393, 25 November 1915, WO339/66665/142133/2, Public Records Office (PRO); Wing Commander Ira Jones, *Tiger Squadron* (London: W.H. Allen/White Lion, 1972), p. 68.
12. Dudgeon adhered to Jones's 1887 date of birth. Dudgeon, *'Mick'*, p. 26.
13. BP would have been a distant figure on the parade ground, but in his unconventional approach to soldiering could easily be a role model for Edward – *Aims of Scouting* declared in 1901 that, 'A scout is a brother to every scout no matter to what class he belongs'. My thanks to Anne Tooke for the suggestion, and the quote.
14. Legge-Pomeroy, *The Story of a Regiment of Horse*, pp. 191, 229–34 and 238–62.
15. J.M. Brereton, *A History of the 4th/7th Royal Dragoon Guards and Their Predecessors* (Catterick: 4th/7th Royal Dragoon Guards, 1982) pp. 291–4. The 19 officers and 539 men of the 7th Dragoon Guards arrived in Canterbury by train from Southampton on 8 August 1904, receiving a civic reception before marching through the city to the barracks; *Kentish Observer*, 11 August 1904.
16. Oughton and Smyth claimed that, as he and Julia could not be divorced because of their religion, Mannock later married bigamously, and fathered more children. Oughton and Smyth, *Ace With One Eye*, p. 41.
17. See Rudyard Kipling, *Kim* (London: Macmillan, 1901). Ironically, Kim's Jesuitical training takes place just about everywhere but inside the walls of St Xavier's.
18. Oughton and Smyth, *Ace With One Eye*, pp. 20–40; Dudgeon, *'Mick'*, pp. 27–30.
19. Regrettably the BBC proved either unable or unwilling to provide details of this 15- or 20-minutes broadcast, including the illustrated booklet which always accompanied such a series. The year of broadcast, coinciding with imminent publication of *Ace With One Eye*, encourages speculation that actor turned author Vernon Smyth may have been involved in one form or another.
20. The Second World War Japanese ace, Saburo Sakai added four more victories to his tally of 60 after recovering from the loss of an eye, and Ira Jones was trained by a 'one-eyed Captain Foggin'. Jones, *King of Air Fighters*, p. 167. My

thanks to Paddy Johnston and Anne Tooke for examples where operational duty was not impaired by the loss of one eye.
21. 'E Mannock', Army Form B.178, 22 May 1915, Army Form M.T. 393, 25 November 1915, and Proceedings of Medical Board, 2 August 1916, WO339/66665/142133/2, PRO.
22. Anonymous family source quoted in Dudgeon, 'Mick', p. 180. Among working-class Edwardian children eye defects were almost the norm, witness the 1902 survey of Edinburgh schools which found 92 per cent of those examined had impaired vision: B. Harris, *The Health of the School Child: a History of the School Medical Service in England and Wales* (Buckingham: Open University Press, 1995) p. 42
23. Jones, *King of Air Fighters*, pp. 19–20; 'McScotch', *Fighter Pilot*, p. 174.
24. The Victorians, including Kipling, equated visual memory with intelligence, and Patrick Mannock's children and grandchildren played Kim's observation game and other tests of instant recall.
25. 'McScotch', *Fighter Pilot*, p. 139; Oughton (ed.), *Personal Diary of Major Edward 'Mick' Mannock*, 14 June 1917, pp. 106–9; Dudgeon, 'Mick', pp. 179–80.
26. Dudgeon, 'Mick', pp. 27–30; Oughton and Smyth, *Ace With One Eye*, pp. 20–40.
27. On Edwardian Canterbury, see Marjorie Lyle, *Canterbury* (London: Batsford/English Heritage, 1994) chapter 7; Michael Winstanley, *Life in Kent at the Turn of the Century* (Folkestone: Dawson, 1978) chapter 12; and above all, *Canterbury 1905* (Canterbury: Parker, 1997), a facsimile edition of Canterbury Chamber of Trade, *The Ancient City of Canterbury* (Canterbury: Cross & Jackman, 1905). The population of the city dropped slightly between the 1901 and 1911 censuses to 24626, of which the garrison made up around 900, hence an identifiable sense of community.
28. The Mannocks moved into 4 Jones Cottages in September 1906. Northgate was the roughest quarter of the city, where 'even the police patrolled in twos'. Winstanley, *Life in Kent At The Turn Of The Century*, p. 165.
29. Equally provocative was the presence from 1880 to 1928 of French Jesuits at St Mary's College, a large and imposing set of buildings in St Stephen's parish, north of the city centre.
30. Thank you to Richard Sexton for checking the school roll, 1890–1912, of St Thomas RC Primary School, Canterbury.
31. The National Telephone Company Ltd's offices were at 39 St George's Street, the southern extension of the High Street nearly obliterated in the 'Baedeker Raid' of 1 June 1942.
32. For all his establishment credentials Gardner was never afraid to ask awkward questions: as soon as he was elected to the Board of Guardians he started questioning long-accepted procedures for administering the city's workhouse; see weekly reports of the Board's meetings in the *Kentish Observer*, commencing 21 April 1910.
33. Information on Cuthbert Gardner (1877–1967) drawn from 'A Canterbury Causerie', *Kentish Observer*, 3 September 1989, reports in the *Kentish Gazette* throughout the 1960s, and 1998 correspondence with G.B. Cotton, a partner in Gardner & Croft, and a former colleague. Although Gardner was Mannock's solicitor as well as his mentor, sadly no papers survive in the practice

archives, nor in the deposits of Gardner and Allen in the Canterbury Cathedral Archives.
34. The Church Lads' Brigade's origins lay in the Victorian Volunteer movement – via secession from the originally non-denominational Boys' Brigade – and by the late 1900s it was affiliated to the King's Royal Rifles Cadets and wore khaki service dress uniform. Anne Summers, 'Militarism in Britain', *History Workshop Journal*, 2 (1976) pp. 119–120.
35. Interview with Frederick Rawson, *Kentish Gazette*, 24 August 1976.
36. Territorial Force weekly orders in *Kentish Gazette*, 1910–13. The paper regularly carried plaintive pleas for fresh recruits, presumably because like so many other Territorial units then and now there was a high turnover of personnel. In the final years of peace the War Office was sensitive to National Service League claims that the Territorial Force had failed to meet its ambitious recruitment targets. Within Whitehall it was common knowledge that, ill equipped and with only 7 per cent prepared to serve overseas, the Territorials could form no part of War Office planning for a BEF. Niall Ferguson, *The Pity of War* (London: Allen Lane, The Penguin Press, 1998) p. 102. On Haldane and the prewar Territorial Force, see Spiers, 'The Late Victorian Army', pp. 206–7, 211. On the National Service League, see Summers, 'Militarism in Britain', pp. 113–17.
37. Profile of A.R. Henchley, *Kentish Gazette*, 26 November 1911.
38. On the distinctly urban character of the Territorial infantry and service battalions – 'a military embodiment of the regions from which they hailed' – see Keegan, *The Face of Battle*, pp. 223–5. Sassoon, billeted outside Canterbury in September 1914, soon discovered that the Yeomanry's training was cursory, and its 'squadron drill was an unsymmetrical affair'. By comparison, Henchley's infantry unit drilled up to three times a week, conducted a monthly route march, ran weekly training courses, and entrained annually to the south coast for a three-week exercise. Siegfried Sassoon, *Memoirs of a Foxhunting Man* (London: Faber & Faber, 1928; paperback edn. 1960) pp. 246–7; Territorial Force weekly orders in *Kentish Gazette*, 1910–13.
39. Gardner found *Ace With One Eye* a good read, but complained that, 'Many of the chapters are written in journalistic form and I have no means of knowing if correct.' Cuthbert Gardner to the Editor, *Kentish Gazette*, 22 May 1964.
40. Henniker Heaton became Canterbury's Member in 1885, readily accepting Lord Halsbury's advice that if questioned on whether he was in favour of paid MPs, 'you must reply that you are in favour of the good old practice of payment for voters and a pension for them afterwards.' He was returned unopposed at the next four elections, and the Unionist poll 1885–1910, including votes cast for the unofficial Bennett-Goldney, averaged 69.3 per cent. Henry Pelling, *Social Geography of British Elections 1885–1910* (London: Macmillan, 1967) p. 81; Neal Blewett, *The Peers, the parties and the People: The General Elections of 1910* (London: Macmillan, 1972) p. 220.
41. To get some idea of a typical crowd, see Derek Butler (ed.), *Canterbury in Old Photographs* (Gloucester: Alan Sutton) p. 17. Imported hops was such a big issue that in May 1908 a demonstration of around 50 000 Kent farmworkers was organized in central London. Free trade was blamed for a sharp drop in commodity prices since the 1870s, reflected in the halving of hop gardens'

acreage by 1914, and an accelerated movement off the land; Winstanley, *Life in Kent at the Turn of the Century*, pp. 18–22

42. The establishment of 'Labour Churches' was a largely Nonconformist initiative which had peaked by 1900, see Alan Wilkinson, *Christian Socialism: Scott Holland to Tony Blair* (London: SCM Press, 1998) pp. 28–9, and Stephen Yeo, 'A New Life: The Religion of Socialism in Britain 1883–1896', *History Workshop Journal*, 4 (1977) pp. 14 and 38–9. Speed lived in one of the recently built terraced houses in Wincheap, a working-class area outside the city walls and adjacent to Canterbury West railway station. At Wincheap Green in early October 1907, 'A good crowd assembled, including in its ranks J.P.s, Tory leaders and tanners.' On Speed's activities, see ILP branch reports, *Labour Leader*, 6, 20 and 27 September 1907, 7 August, 11 September, and 9 and 30 October 1908. The depth of hostility to the pro-free trade ILP is highlighted by the contrasting experience of Mrs Pankhurst when she addressed a WSPU rally in April 1910 – 'a very civilised occasion' with no heckling according to the *Kentish Observer*, 28 April 1910.
43. Obituary of Walter Speed, *Labour Leader*, 3 October 1911. Speed, 28 when he died, had been employed (by Ben Keeling?) in the Leeds Labour Exchange after six months exhausting work organizing the Yorkshire ILP Divisional Council. My thanks to Fred Whitemore for allowing me to draw upon his research into the Canterbury ILP.
44. 'Hero who ruled the skies', *Kentish Gazette*, 21 July 1978. The Church Lads' Brigade was closely linked to the temperance movement, thereby reinforcing Mannock's indifference to drink.
45. Oughton and Smyth mention a brief pubescent infatuation with one Grace Wimsett, but there is no mention of any other women in Canterbury. Oughton and Smyth, *Ace With One Eye*, pp. 66–9.
46. To compound Julia's woes, Patrick got married in an Anglican church in 1915.
47. Information received with thanks from D.B. Pratt, Wellingborough, re his aunt's acquaintance with Mannock.
48. On the size and sophistication of the telegraph and telephone system by the 1900s, see Tom Standage, *The Victorian Internet: the remarkable story of the telegraph and the nineteenth century's online pioneers* (London: Weidenfeld & Nicolson, 1998).
49. Robert Tressell was still working on the final draft of his novel when he died, just as Mannock began working in Wellingborough. Frank Owen, the tubercular socialist hero who paints for Rushton & Co., has precious few comrades in Mugsborough [Hastings], although the town parallels Wellingborough in that the 'Brigands' who control the council are Liberal 'well-to-do tradesmen...the fact that a man had succeeded in accumulating money in business was a clear demonstration of his fitness to be entrusted with the business of the town.' Robert Tressell, *The Ragged Trousered Philanthropists* (London: Panther, 1965) p. 192.
50. 'McScotch', *Fighter Pilot*, p.13; Ira Jones to Vernon Smyth, 1 May 1957, Edward Mannock papers, Misc. 117 Item 1839, Imperial War Museum (IWM).
51. Information received with thanks from Lyndon Garfirth, Wellingborough.
52. By 1897 Wellingborough was producing 33 950 tons per year. Joyce and Maurice Palmer, *A History of Wellingborough* (Earls Barton: Steepleprint, 1972) p. 217. On the impact of industrialization on late nineteenth-century

Wellingborough, see also Joyce and Maurice Palmer, *Wellingborough Memories* (Wellingborough: W.D. Wharton, 1995).
53. Founding the Liberal Club in 1900, a disastrous year for the party in terms of internal bickering and electoral defeat, suggests a laudable degree of optimism.
54. John Bunyan was born down the road in Bedford. Northampton had a healthy tolerance of secularism, one of its two Liberal MPs in the 1880s being Charles Bradlaugh, famous for refusing to take a religious oath on entering the Commons. His legacy was a distinctly secular brand of socialism in Edwardian Northampton, rooted in the Social Democratic Federation/ Party (SDP). This was very different from the Wellingborough ILP, led by churchgoers like Eyles and Mannock. Pelling, *Social Geography of British Elections 1885–1910*, p. 110.
55. Palmer, *A History of Wellingborough*, p. 195; A. Fox, *History of the National Union of Boot and Shoe Operatives* (Oxford: Oxford University Press) pp. 356–7.
56. Duncan Tanner, *Political Change and the Labour party, 1900–1918* (Cambridge: Cambridge University Press, 1990) pp. 297–8. On prewar relations between Labour and the Co-operative Union, see Ross McKibbin, *The Evolution of the Labour Party* (Oxford: Oxford University Press, 1974) pp. 43–7.
57. Pelling, *Social Geography of British Elections 1885–1910*, pp. 114–15. Both Northampton and Northamptonshire East were notable for the Liberals' success in revitalizing the party machine and recruiting new members, making Labour's modest inroad into Chiozza Money's majority that much more impressive. Tanner, *Political Change and the Labour party, 1900–1918*, pp. 289–90.
58. For the size and influence of the Fabian Society relative to the ILP, see Adrian Smith, *The New Statesman Portrait of a Political Weekly 1913–1931* (London: Frank Cass, 1996) pp. 18–19.
59. By 1911 the 'Big Four' were in practice five, Arthur Henderson having served as chairman of the parliamentary party in 1908–9 before becoming chief whip and ILP treasurer.
60. For all the purple prose MacDonald was as persuasive in print as in person – witness his prolific newspaper articles and prewar contribution to the Socialist Library series, which he edited: *Socialism and Society* (London: ILP, 1905) and the two volumes of *Socialism and Government* (London: ILP, 1909). R.E. Dowse, *Left In The Centre: the Independent Labour Party 1893–1940* (London: Longman, 1966) pp. 6–7; Paul Ward, *Red Flag and Union Jack Englishness, Patriotism and the British Left, 1881–1924* (Woodbridge: Boydell Press/Royal Historical Society, 1998) pp. 84–5. Stephen Yeo has argued that the intense engagement and the keen sense of socialist crusade that was such a marked feature of the early Labour movement had peaked by the late 1890s. However, for much of the Edwardian period there remained an absence of disillusion and opportunism at grassroots level, especially in relatively new ILP branches such as Wellingborough. Stephen Yeo, 'A New Life: The Religion of Socialism in Britain 1883–1896', pp. 3–56.
61. Ward, *Red Flag and Union Jack*, p. 93. While the keenest activist would have read the *Labour Leader* most ILP members contented themselves with the *ILP News* and the 70 or so regional weeklies that appeared regularly or intermittently between 1893 and 1914. Robert Blatchford's idiosyncratic weekly, the

Clarion, remained popular despite its editor's increasingly eccentric and jingoistic view of the world. Deian Hopkin, 'The Labour Party Press' in K.D. Brown (ed.), *The First Labour Party 1906–14* (London: Croom Helm, 1985) pp. 105–28.
62. The Eyles family lived at 183 Mill Road until 1924 when they moved to 195. Derek Eyles was killed during the Second World War, and his father died in 1959.
63. Ideal Clothiers Ltd opened a large factory in 1900, providing a major boost to local women's chances of securing a job. The shoemakers Rudlen Ltd survived a serious fire in 1913, and, like Ideal Clothiers, prospered during the First World War. Standing outside the factory in April 1999 I realized that this was where my shoes had been made – Dr Martens originate in Mill Road.
64. Jim Eyles quoted in Dudgeon, '*Mick*', p. 35.
65. Ibid.
66. Ibid.
67. George Dangerfield, *The Strange Death of Liberal England* (London: Paladin, 1970). Largely discredited today, Dangerfield's stylish, semi-apocalyptic vision of a Britain rent asunder by suffragettes, syndicalists and warring Irishmen, but saved from itself by Sarajevo, first appeared in 1935. Jim Eyles quoted in Dudgeon, '*Mick*', p. 37.
68. On the importance of autodidactism within the early Labour movement, see Ross McKibbin, 'Why was there no Marxism in Britain?', *The Ideologies of Class: Social Relations in Britain 1880–1950* (Oxford: Oxford University Press, 1990) pp. 34–5.
69. Dudgeon wrongly claimed that in London Mabel Eyles introduced him to Ivor Novello's mother, who was sufficiently impressed that she gave him a few singing lessons. If this did happen then the contact was Dorothy Mannock, Patrick's wife. Dudgeon, '*Mick*', pp. 35–6; family information. Mannock must have been technically accomplished on the violin as according to the family his favourite piece was Fritz Kreisler's demanding 'Caprice Viennois'. The latter was clearly an RFC favourite as Cecil Lewis mentions it playing on the phonograph on more sombre nights in the mess. Lewis, *Sagittarius Rising*, p. 227.
70. Jim Eyles quoted in Dudgeon, '*Mick*', p. 37, and Jones, *King of Air Fighters*, pp. 8–9.
71. Oughton and Smyth, *Ace With One Eye*, pp.107–8.
72. Norwich-born W.R. Smith lost his Wellingborough seat in 1922, but represented his home town for the duration of the first two Labour governments, in both of which he served as a parliamentary secretary.
73. Dowse, *Left In The Centre*, pp. 10, 18–19 and 29; McKibbin, *The Evolution of the Labour Party*, pp. 19, 33, and 47.
74. National Administrative Council, 'Hints for I.L.P. Secretaries', p. 7, ILP5 Box 9, 1913/19, Independent Labour Party Papers, British Library of Political and Economic Science.
75. Jim Eyles quoted in Jones, *King of Air Fighters*, pp. 8–9. Sadly, the records of the Wellingborough ILP have long since disappeared, the same being true of the Eyles family papers. Oughton and Smyth hint at Mannock keeping a diary, but efforts by both his niece and myself to track down his diaries proved fruitless. Thanks to all those in Wellingborough and beyond who assisted in

the abortive search for prewar papers, especially Dorothy Knight of the Wellingborough Constituency Labour Party.
76. Ibid. Predictably, this becomes a splendid story of naked class war in Oughton and Smyth, *Ace With One Eye*, pp. 79–80.
77. Jim Eyles quoted in Jones, *King of Air Fighters*, p. 8; information received with thanks from Lyndon Garfirth, Wellingborough.
78. Clifford Maycock, *The Parish Church of St Mary the Virgin Wellingborough*, and *Notes on St Mary the Virgin Wellingborough*, 1 and 2, church pamphlets, no date.
79. Individual statements of belief and mission statement of St Mary's, Wellingborough. The present priest is in fact a member of the Society of the Holy Cross, and uses current Roman Catholic liturgy.
80. Albeit the only name with decorations also listed.
81. Until Cardinal Bourne concluded in 1924 that Labour was not really a socialist party, and thus not hostile to religion, the bishops – if not necessarily those priests in closer contact with their Labour-voting working-class flock – discouraged membership. The Catholic Social Guild (1909–67) had less than 4000 members at its peak. Wheatley's tiny Catholic Socialist Society was based on his native Glasgow. Wilkinson, *Christian Socialism*, pp. 36–7, 89–90, 148, 151 and 213–15.
82. Ibid., pp. 9–11, 100, 108 and 141. Edward Norman, *Church and Society in England 1770–1970* (Oxford: Oxford University Press, 1976) p. 317.
83. Stewart Headlam's 1907 *The Socialist's Church* quoted in Wilkinson, *Christian Socialism*, p. 36.
84. Ibid., pp. 66 and 134. Yeo, 'A New Life: The Religion of Socialism in Britain 1883–1896', pp. 7 and 18.
85. Ibid., p. 23.
86. Summers, 'Militarism in Britain before the Great War', pp. 118 and 119.
87. Indeed, H.M. Hyndman's Social Democratic Federation argued for compulsory service and the end of a standing army. Robert Blatchford's *Clarion, Justice* and the *New Age* all put the case for a citizen army and a big navy, the first two titles articulating a minority of British socialists' deep-rooted suspicion of Germany. Ward, *Red Flag and Union Jack*, pp. 108–13.
88. See Tawney's recollection of the war in *The Acquisitive Society* (London: G. Bell & Sons Ltd, 1921) and Wilkinson, *Christian Socialism*, pp. 102–3.
89. Ward, *Red Flag and Union Jack*, pp. 106–7 and 5.
90. J. Ramsay MacDonald, *Socialism* (London: ILP, 1907) p. 117, quoted in ibid., p. 50. The increasingly Germanophobe Hyndman encouraged the notion of an 'un-British' ideology by rarely acknowledging that the SDF was – at least in the beginning – a Marxist party.
91. Ibid., p. 5.
92. Geoffrey Field quoted in ibid., p. 4.
93. Tanner, *Political Change and the Labour party, 1900–1918*, pp. 301, 311, 333 and 413. On the ILP's ability after August 1914 to reconcile pro- and anti-war factions, see Dowse, *Left In The Centre*, pp. 20–3. Although in August 1914 only two of the ILP's MPs supported the war, and most branches endorsed the NAC's anti-war manifesto, Fenner Brockway estimated over one-fifth of the membership were openly pro-war – the leadership sought to accommodate, not censure, the minority. Ward, *Red Flag and Union Jack*, pp. 126–7.

Chapter 3 Preparing for War, 1914–17

1. Passport issued to Edward Mannock, 10 January 1914, Foreign Office. AC 72/10 1, Aviation Records Department, RAF Museum (RAFM). Dudgeon claimed Mannock's walking stick was hollow and held gold sovereigns given to him by friends in Wellingborough. The passport photograph certainly suggests a sense of wellbeing and prosperity. Dudgeon, *'Mick'*, p. 38.
2. Jim Eyles quoted in ibid., pp. 37–8.
3. Ibid.: 'When he walked out of our little home...I knew instinctively that he was off to something great.'
4. Letter in transit from Edward Mannock to Jim and Mabel Eyles, spring 1914, quoted in Oughton and Smyth, *Ace With One Eye*, pp. 83–4.
5. Summary of Anglo-Ottoman relations by 1914 based upon Joseph Heller, *British Policy Towards the Ottoman Empire* (London: Frank Cass, 1983) pp. 1–7, 116, 120, 130 and 133, and Robert Rhodes James, *Gallipoli* (London: Batsford, 1965; Pimlico, 1999), pp. 4–8.
6. One of Mannock's 'chiefs' was M. Webster Jenkinson who, while not a spy as such, passed on general impressions of the political situation to the Foreign Office. His intelligence was poor – witness his writing an upbeat assessment of prospects for continued neutrality at the very moment the Ottoman navy was shelling Russian ports. M. Webster Jenkinson to Sir Edward Grey, 30 October 1914, FO371/2145/65657. 270, PRO, now the National Archives.
7. Letters to Jim and Mabel Eyles, and 1914 diary, quoted and drawn on in Oughton and Smyth, *Ace With One Eye*, pp. 85–93, and Dudgeon, *'Mick'*, pp. 38–40.
8. Edward Mannock to Jim Eyles, 10 July 1914, quoted in Jones, *King of Air Fighters*, p. 15.
9. Ibid. Mannock wore a signet ring on his marriage finger: 'Stiff upper lip, as becomes a rising engineer.'
10. Dudgeon, *'Mick'*, pp. 38–40.
11. Heller, *British Policy Towards the Ottoman Empire*, pp. 134, 138–9 and 145–6; Rhodes James, *Gallipoli*, pp. 9–10. See also A.L. Macfie, *The End of the Ottoman Empire, 1908–1923* (London: Longman, 1998). Letters to Jim and Mabel Eyles, and to Julia Mannock, quoted and drawn on in Oughton and Smyth, *Ace With One Eye*, pp. 94–7. Edward Mannock to Jim Eyles, 28 August 1914, quoted in Jones, *King of Air Fighters*, p. 16.
12. Agreement to earlier request by HMG of US protection of British non-combatants in Turkey, Walter Hines Page (US Ambassador in London) to Sir Edward Grey, 6 November 1914, FO371/2145/65657 195, PRO; Heller, *British Policy Towards the Ottoman Empire*, p. 140, 147–9. Morgenthau was the same diplomat and prominent 'New Dealer' responsible for the abortive 1944 plan to strip a defeated Germany of both its industrial capacity and its territorial integrity.
13. Indeed the British community increasingly complained of embassy indifference, a charge vehemently denied by an ambassador increasingly alarmed by a lack of urgency in leaving Constantinople. Sir Louis Mallet to Sir Edwin Pears, 20 October 1914, Cornelius Van H. Engert papers, Box 12 Folder 4, Lauinger Library, Georgetown University (LIGU).

14. Edward Mannock to Mabel Eyles, 21 October 1914, quoted in Jones, *King of Air Fighters*, p. 16.
15. Heller, *British Policy Towards the Ottoman Empire*, pp. 147, 149 and 151–4; Rhodes James, *Gallipoli*, pp. 12–13.
16. F. Elliott (British Embassy in Athens) to Sir Edward Grey, 12 November 1914, FO3712146/258, PRO.
17. Memo from US Embassy to HM Under-Secretary of State for Foreign Affairs, replicating telegram from Henry Morgenthau of 6 November 1914, 13 November 1914, FO3712145/71065 214, PRO.
18. Eyre Crowe (Foreign Office) to Home, India and Colonial Offices, 7 November 1914, FO371/2145/6636, PRO.
19. See files in 'Turkey – treatment of foreigners and property 1915', FO371/1244; Walter Hines Page to Sir Edward Grey, 13 and 15 January 1914, FO371/2481/4573 and 5449, PRO.
20. Sir Edward Grey to Walter Hines Page, 7 May 1915, FO371/2489/54163, PRO. Fifty-one men aged 20 to 40 with French or British passports were briefly sent by steamer to Gallipoli, but interestingly only four were actually born in their respective countries. Walter Hines Page to Sir Edward Grey, 8 May 1915, FO371/2487/56708, PRO. Another reason for the Foreign Office's uncompromising position was that, as PRO files on Turkey readily confirm, officials clearly shared the optimism of the Admiralty and later the War Office that the Gallipoli offensive would ultimately be successful – witness detailed arrangements for the future international administration of Constantinople (not just which countries would run what, but who specifically).
21. See files in 'Turkey – detention of civilians 1915' and 'British subjects in Turkey 1915', FO371/1244, PRO.
22. Secretary of US Embassy Constantinople, reply to enquiry by Jim Eyles of 11 January 1915, 19 February 1915, quoted in Jones, *King of Air Fighters*, p. 17. Eyles had received only one letter in over ten weeks, hence his inquiry.
23. 'If you were to ask for the reason of this stupendous collapse of a whole nation, one single word might almost suffice as an explanation, namely: *inefficiency*.' Cornelius Van H. Engert to President (University of California at Berkeley?) Lowell, 5 December 1912, Cornelius Van H. Engert papers, Box 1 Folder 33, LIGU. My thanks to Derek Edgell and Alan Brown for IT advice on tracing members of the wartime American Embassy.
24. Cornelius Van H. Engert to Del (?), 10 March 1915, and to (?) Burdett, 14 March 1915, ibid. As vice-consul at Chanak in December 1914 Engert had by chance witnessed a RN submarine torpedo the Ottoman battleship *Messudieh*, and been much impressed. Rhodes James, *Gallipoli*, p. 13.
25. Oughton and Smyth, *Ace With One Eye*, p. 104. 'E Mannock', Army Form B. 178, 22 May 1915, WO339/66665/142133, PRO.
26. The medical officer judged Mannock's overall physical condition 'good', and his weight of 153 lb matched his height of 5' 10". Ibid.
27. Ibid., pp. 97–104; Dudgeon '*Mick*', pp. 42–6.
28. Florence Minter quoted in Jones, *King of Air Fighters*, p. 18.
29. Oughton and Smyth, *Ace With One Eye*, p. 102; Dudgeon '*Mick*', pp. 44–5; Jones, *King of Air Fighters*, p. 18.
30. Edward Mannock to Mabel Eyles, 21 October 1914, quoted in Jones, *King of Air Fighters*, p. 16.

31. Brought up in a paternalistic empire, Mannock's personal indignation would have been complemented by outrage at the Turks' appalling treatment of the incarcerated Indians, all of whom clung on to the belief that 'the English will come'. Ibid., pp. 18 and 36.
32. Jim Eyles quoted in Dudgeon, *'Mick'*, p. 47.
33. Territorial soldiers 1914–16 still had to be asked whether they were willing to serve overseas (before the war only 7 per cent had said yes). Ferguson, *The Pity of War*, p. 102.
34. 'Attestation of Edward Mannock to serve 4 years in the Territorial Force', 1915–16, WO339/66665/142133/3, PRO.
35. Anonymous RAMC sergeant (R. Wyles) quoted in Jones, *King of Air Fighters*, p. 20.
36. Thirteen per cent of the hugely popular *Battle of the Somme*'s 77 minutes focused upon the dead and wounded, and in the final quarter of the film over 40 per cent. Released in August 1916, up to twenty million cinema-goers saw the film in its first six weeks. Nicholas Reeves, 'The Real Thing At Last: *Battle of the Somme* and the domestic cinema audience in the autumn of 1916', *The Historian*, 51 (1996) pp. 4–8.
37. Anonymous RAMC sergeant (R. Wyles) quoted in Jones, *King of Air Fighters*, p. 21. R. Wyles to Vernon Smyth, 9 April 1957, Edward Mannock papers, Misc. 117 Item 1839, IWM. Wyles, another RAMC sergeant and future officer, served with Mannock from 1910. While at Halton Park 'Jerry' agreed to be his best man.
38. Oughton and Smyth, *Ace With One Eye*, p. 111; Dowse, *Left In The Centre*, pp. 29 and 23. In contrast, when R.H. Tawney met three leading Fabians – W.S. Sanders, Mostyn Lloyd, and 'Ben' Keeling at St Omer early in 1916 – they drafted a manifesto from serving trade unionists to striking Clydeside munitions workers. J.M. Winter, *Socialism and the Challenge of War: Ideas and Politics in Britain, 1912–18* (London: Routledge & Kegan Paul, 1974) p. 151.
39. For the continuing debate on the political influence of ABCA, see J.A. Crang, 'Army Education and the 1945 General Election', *History*, LXXXI (1996) pp. 215–27.
40. Anonymous RAMC sergeant (R. Wyles) quoted in Jones, *King of Air Fighters*, p. 21; 'Major Mannock of the R.A.F. By An Old Comrade' (R. Wyles), *Daily Mail*, (?) 1918 – cutting in Edward Mannock papers, IWM; Oughton, 'Biographical sketch of Mick Mannock', *The Personal Diary of 'Mick' Mannock*, p. 17.
41. Ibid.
42. There is no War Office record of military intelligence being approached, nor any reference to the mock parliament in RAMC records. On the other hand it is hard to believe that the mock parliament was not under close surveillance by officers at the camp. My thanks to the PRO and the RAMC Museum for searching for archival evidence of surveillance.
43. National Council for Civil Liberties (Captain Gilbert Hall), *The Story of the Cairo Forces' Parliament* (London: NCCL, 1944); (Sergeant) Bill Davidson, 'Essay in Labour History: the Cairo Forces Parliament', *Labour History Review*, LV (1990), pp. 20–6. Leo Abse doubled the numbers in attendance when interviewed for *Now The War Is Over*, BBC2, 1985, episode 1.
44. David Englander and James Osborne, 'Jack, Tommy and Henry Dubb: the Armed Forces and the Working Class', *The Historical Journal*, XXI (1978), pp. 602–5.

45. Anonymous RAMC sergeant (R. Wyles) quoted in Jones, *King of Air Fighters*, p. 21.
46. Ibid.
47. Anonymous RAMC sergeant (R. Wyles) quoted in Jones, *King of Air Fighters*, p. 21; 'Application of Edward Mannock for appointment to a temporary commission in the regular army for the period of the war', 25 November 1915, WO339/66665/142133, PRO.
48. Ibid.; Edward Mannock to Jim Eyles, (?) November 1915, quoted in Oughton and Smyth, *Ace With One Eye*, p. 110.
49. 'Application of Edward Mannock for appointment to a temporary commission in the regular army for the period of the war', 25 November 1915, WO339/66665/142133, PRO; *Statistics of the British Empire During the Great War* (London: HMSO, 1922) p. 707.
50. Before joining his first squadron Mannock drank very little if at all. A fellow trainee pilot assumed, 'he was a staunch teetotaller'. Captain (later Wing Commander) Meredith Thompson, quoted in Jones, *King of Air Fighters*, p. 35.
51. Mannock's experience of Fenny Stratford based on the testimony of Lieutenant J. E. Buchanan, assistant adjutant, reproduced in Jones, *King of Air Fighters*, pp. 22–5.
52. Ibid., p. 24.
53. Ibid.
54. Francis Ledwidge quoted in Angus Calder (ed.), *Wars* (London: Penguin, 1999) p. 109.
55. Jane Leonard, 'The Reaction of Irish Officers in the British Army to the Easter Rising of 1916', in Hugh Cecil and Peter H. Liddle (eds), *Facing Armageddon: The First World War Experienced* (London: Leo Cooper, 1996) p. 265; W.B. Yeats, 'Easter 1916'.
56. Lieutenant J.E. Buchanan, in Jones, *King of Air Fighters*, p. 23. The British Army has a tradition of promotion from the ranks entailing basic training alongside newly selected officer cadets. When teaching at Sandhurst I never failed to be impressed by the patience and restraint of experienced soldiers faced with an initial six weeks of ceaseless drill and bull.
57. Ibid.
58. Ibid.
59. Ibid., p. 22. The wonderfully named Francis Lucins Pym Mannock, better known as plain Frank, was the eldest son of Edward Corringham/Mannock's second eldest brother. His names alone indicate his metropolitan middle-class background. Smyth claimed Buchanan also knew Frank's younger brother, Charles, and billiards champion John Mannock's son, Patrick, who started reviewing films and plays for the *Daily Herald* in the 1930s and was still a reasonably well known journalist 20 years later. Oughton and Smyth, *Ace With One Eye*, p. 113.
60. Lieutenant J.E. Buchanan, in Jones, *King of Air Fighters*, p. 24–5.
61. Ibid., p. 125–7; Jones, *King of Air Fighters*, p. 26–7. Jones reassured the doctor in Bedford that he 'need suffer no remorse' for being deceived by...a determined man...overcoming obstacles in order to be at the front, rubbing shoulders with MEN WHO *were* MEN, and helping to kill the enemy regardless of his own life.' Ibid., pp. 27 and 26.

62. 'E Mannock – Proceedings of Medical Board', 2 August 1916, and medical report, 24 November 1915, WO339/6665/142133, PRO.
63. Lieutenant J.E. Buchanan, in Jones, *King of Air Fighters*, p. 24.
64. Ibid., p. 25.
65. Oughton and Smyth, *Ace With One Eye*, pp. 116–19.
66. Ray Sturtivant, 'British Flying Training in World War I', *Cross and Cockade*, XXV (1994) pp. 18–20.
67. In the course of the war around 8000 pilots died in training, while 6166 were killed in action.
68. Albert Ball to Albert Ball (senior), 1 July 1916, Correspondence of Captain Albert Ball, Nottinghamshire Record Office (NRO). My thanks to Professor Bob White for passing on to me his research notes on Ball's papers.
69. Steel and Hart, *Tumult In The Clouds*, p. 125; Norman Franks, *Who Downed the Aces in WWI?* (London: Grub Street, 1996) p. 1 and 23. Increased newspaper coverage by both sides in 1916 encouraged a more systematic method of categorizing 'kills'. The Allies, fighting an offensive war over the German lines, had to develop recognized criteria for confirming victories, for details of which see ibid., pp. 24–5.
70. Ibid., p. 1.
71. Chaz Bowyer, *Albert Ball, VC* (Wrexham: Bridge Books, 1994); Barker, *The Royal Flying Corps In France*, pp. 40–2.
72. 9 May 1917, Oughton (ed.), *The Personal Diary of 'Mick' Mannock*, pp. 69–71.
73. Again my thanks to Professor Bob White, for pointing out that Ball shared Mannock and McCudden's enthusiasm for tinkering with machinery.
74. August 1916 entry in lost diary, quoted in Jones, *King of Air Fighters*, p. 27.
75. Sturtivant, 'British Flying Training in World War I', pp. 20–1; Steel and Hart, *Tumult In The Clouds*, pp. 84–7.
76. Chronicle of Mannock's initial training based on Jones, *King Of Air Fighters*, p. 31, and Sturtivant, 'British Flying Training in World War I', pp. 18–21.
77. Chaz Bowyer (ed.), Gwilym H. Lewis, DFC, *Wings Over The Somme 1916–18* (Wrexham: Bridge Books, 1976; 1996 edn) pp. 15–16.
78. Training was now formalized, with manuals and regulations, for an amusing example of which see H. Clarke (ed.), 'Rules of the Air and Hints on Flying: Instructions for Students of the CFS Detachment at RFC Netheravon', *Cross and Cockade*, XXIII (1992) pp. 214–16.
79. Oughton and Smyth, *Ace With One Eye*, pp. 147–8.
80. 'McScotch', *Fighter Pilot*, p. 218.
81. Residents of Mill Road can name deceased neighbours who insisted on having seen Mannock's aeroplane parked at Wellingborough School.
82. Steel and Hart, *Tumult In The Clouds*, pp. 87–8.
83. Captain Chapman, quoted in Jones, *King of Air Fighters*, p. 32.
84. 'McScotch', *Fighter Pilot*, pp. 223–4.
85. Oughton and Smyth, *Ace With One Eye*, pp. 128–38.
86. Captain (later Wing Commander) Meredith Thompson, quoted in Jones, *King of Air Fighters*, p. 36.
87. 'James McCudden', VC Box 31, IWM; Peter G. Cooksly, 'J.T.B. McCudden', *VCs of the First World War: The Air Aces* (Stroud: Sutton Publishing, 1996) pp. 136–50. McCudden had two brothers in the RFC: his elder brother, Bill, died when his monoplane broke up on landing in April 1915, while Lieutenant

John McCudden VC was shot down over Busigny in March 1918. A third brother served in the Royal Navy and was killed in May 1915.
88. C.G. Grey (ed.), James Thomas Byford McCudden, *Five Years In The Royal Flying Corps* (London: The Aeroplane & General Publishing Co. Ltd, 1918) p. 174. McCudden's wartime memoirs were reissued between the wars as *Flying Fury* (London: John Hamilton, 1930).
89. Ibid.; Captain (later Wing Commander) Meredith Thompson, quoted in Jones, *King of Air Fighters*, p. 36. Oughton and Smyth, *Ace With One Eye*, pp. 140 and 145.
90. Oughton and Smyth, *Ace With One Eye*, pp. 140–4; Dudgeon, '*Mick*', pp. 55–6.
91. McCudden, *Five Years In The Royal Flying Corps*.
92. Captain (later Wing Commander) Meredith Thompson, quoted in Jones, *King of Air Fighters*, p. 36.
93. 1–7 April 1917, Oughton (ed.), *The Personal Diary of 'Mick' Mannock*, pp. 26–33; Edward Mannock to Jim Eyles, 3 April 1917, quoted in Jones, *King of Air Fighters*, p. 66. Mannock had agreed a simple code with Eyles to circumvent the censor and make clear whereabouts he was stationed: the first two letters of the words in the first and second sentence. Also, a request for boots meant he was about to move from his current location. Ibid., pp. 66–7.
94. Ibid.; Barker, *The Royal Flying Corps in France*, p. 30.

Chapter 4   40 Squadron, 1917–18

1. The RFC's loss rate for aircraft peaked at 235, as late as September 1918.
2. The Allies had a 20 per cent superiority in aircraft production by 1918. Ferguson, *The Pity of War*, p. 260.
3. Prewar Germany enjoyed a near monopoly of magneto production. The absence of volume aero-engine production in Britain at the start of the war – plus the absence of an effective procurement system, even after the Ministry of Munitions assumed responsibility in December 1916 – meant a heavy dependence upon French manufacturers until late 1917.
4. Yeates, *Winged Victory*, pp. 146–7. The long and detailed descriptions of Camel pilots' tactics are one reason why critics considered *Winged Victory* such a dull novel. Henry Williamson, 'Tribute to V.M. Yeates', ibid., p. 5. For comparison of the SE5a and the Sopwith Camel, see Hart and Steel, *Tumult In The Clouds*, pp. 221–2.
5. In fiction, e.g. Derek Robinson, *Goshawk Squadron* (London: Heinemann, 1971; HarperCollins, 1993), especially the black humour of chapter 1; and Yeates, *Winged Victory*, where Tom Cundall's advice to new pilots is well-intentioned but amounts to little more than 'keep your eyes open and learn fast'.
6. Alan Clark, 'The fighting spirit', *Daily Telegraph*, 19 September 1999; Steel and Hart, *Tumult In The Clouds*, p. 299.
7. Captain Todd quoted in Jones, *King of Air Fighters*, p. 72, and Captain Henry Jaffe quoted in Oughton and Smyth, *Ace With One Eye*, p. 151. Mannock clearly failed to appreciate the poor impression he had made on the 'good fellows' in the mess; see Edward Mannock to Jim and Mabel Eyles, and to Jessie Mannock, 7 April 1917, each quoted in Jones, *King of Air Fighters*, p. 73–4. He still addressed the Eyles family as 'Dear Comrades', and signed off 'Sincerely and Socialistically Yours'.

8. Lieutenant Lionel Blaxland quoted in Dudgeon, 'Mick', p. 64.
9. Robert Graves, Goodbye to All That (London: Jonathan Cape, 1929; Penguin, 1960) p. 157.
10. Squadron Leader (?) de Burgh, quoted in Jones, King of Air Fighters, pp. 101–2. De Burgh's favourable impression of Mannock was shared with W. A. Bond, a former journalist on The Times whose letters home – to An Airman's Wife – were posthumously published after the war.
11. 14 June 1917, ibid., p. 111.
12. 'If one of the team in Rugger is seriously hurt, he is carried sympathetically off the field, but the game carries on just the same. This spirit of "Carry on" found its counterpart only to a greater degree in the R.F.C. spirit.' Jones, King of Air Fighters, pp. 76 and 75.
13. 13 April 1917, Oughton (ed.), The Personal Diary of 'Mick' Mannock, p. 41. Dudgeon, 'Mick', p. 66; Jones, King of Air Fighters, p. 77.
14. Captain F.L. Barwell, 29 April 1917, over Lens.
15. Air Vice Marshal Arthur Gould Lee quoted in Clark, 'The fighting spirit'.
16. 20 April 1917, Oughton (ed.), The Personal Diary of 'Mick' Mannock p. 49.
17. Edward Mannock to Jim Eyles, 16 April 1917, and to Julia Mannock, 19 and 27 April 1917, ibid., pp. 77–8, 79–80 and 86–7.
18. 20 April, 3 and 9 May 1917, and recollection of Sergeant W. Bovett, ibid., pp. 45–9, 57–9, 69–77 and 153 footnote 31.
19. Captain G.L. Lloyd quoted in Jones, King of Air Fighters, pp. 114–15; Dudgeon, 'Mick', pp. 66–7.
20. Sergeant W. Bovett and Major J. L. T. Pearce quoted in Jones, King of Air Fighters, pp. 82 and 130–3; Oughton (ed.), The Personal Diary of 'Mick' Mannock, p. 153 footnote 31; Steel and Hart, Tumult In The Clouds, p. 242.
21. 7 June 1917, ibid., pp. 103–5 and 153 footnote 31; Steel and Hart, Tumult In The Clouds, pp. 216 and 242; Dudgeon, 'Mick', pp. 68–9; Keith L. Caldwell to Vernon Smyth, Edward Mannock papers, Misc. 117 Item 1839, IWM.
22. Edward Mannock to Mabel Eyles, (?) June 1917, quoted in Jones, King of Air Fighters, p. 134; Steel and Hart, Tumult In The Clouds, pp. 322–5.
23. Edward Mannock to Jim and Mabel Eyles, 12 May 1917, in Jones, King of Air Fighters, p. 116; ibid., pp. 118–19.
24. Clark, 'The fighting spirit': 'Cowardice was a deadly sin.'
25. 9 May 1917, Oughton (ed.), The Personal Diary of 'Mick' Mannock, pp. 70–1. The balloon-strafing was sufficiently important to secure congratulations from Haig and be reported to Lloyd George; see OC 10th (Army) Wing to OC No. 40 Squadron, 13 May 1917, replicated in ibid. (no pagination).
26. Albert Ball to Albert Ball (senior), 5 May 1917, Correspondence of Captain Albert Ball, NRO.
27. 9 May 1917, Oughton (ed.), The Personal Diary of 'Mick' Mannock, pp. 71–5.
28. 14 May 1917, ibid., pp. 83–5. Dudgeon, 'Mick', pp. 74–5.
29. Lionel Blaxland to Julia Mannock, 18 May 1917, in Jones, King of Air Fighters, pp.103–4.
30. 18 and 25 May 1917, Oughton (ed.), The Personal Diary of 'Mick' Mannock, pp. 85–7 and 97; 'McScotch', Fighter Pilot, p. 146. Keymer died of pleurisy in 1924, having served four years at Cranwell. Two of his seven children died flying with the RAF in the Second World War. Bowyer (ed.), Wings Over The Somme

1916–18, p. 121; W.A. Bond's profile of Keymer in Jones, *King of Air Fighters*, p. 98.
31. 'McScotch', *Fighter Pilot*, p. 57.
32. Using the RFC's official categories, both Mannock's victims were 'driven down out of control', as opposed to 'driven down', or the more easily verifiable 'destroyed'. Results in any of the three categories had to be confirmed, except when a pilot's word was accepted for a victory secured while on lone patrol. The latter privilege fuelled rumours concerning certain pilots who were officers but not always considered gentlemen. Jones, *King of Air Fighters*, pp. 124–5; 25 May and 2, 7 and 14 June 1917, Oughton (ed.), *The Personal Diary of 'Mick' Mannock*, pp. 93–107.
33. 14 and 16 June 1917, ibid., pp. 107–13.
34. Ibid., 31 August 1917, p. 133. The incident actually occurred on 19 August 1917. 'McScotch', *Fighter Pilot*, pp. 139–40 and 173–4.
35. Jones seems to imply that Jess was with her mother when he quotes her as saying that Edward never said 'good-bye' but 'au revoir' when returning to France. Jones, *King of Air Fighters*, p. 133.
36. Jim Eyles quoted in Dudgeon, '*Mick*', p. 88; Edward Mannock to Mabel Eyles (?) June 1917, Oughton (ed.), *The Personal Diary of 'Mick' Mannock*, p. 163 footnote 94.
37. The general visited 40 Squadron to congratulate Tilney's men on their innovative balloon-strafing tactics, and was followed in July by no less a person than George V. Cecil Lewis described in very different terms from Mannock a typical squadron visit by Trenchard. 2 June and 20 July 1917, ibid., pp. 103 and 115, and 163 footnotes 94 and 97. Jones, *King of Air Fighters*, p. 132. Lewis, *Sagittarius Fighting*, p. 101.
38. Edward Mannock to Patrick Mannock, 8 May 1918, AC 72/10 11, Aviation Records Department, RAFM; Patrick Mannock quoted in Jones, *King of Air Fighters*, p. 133; 'The Growing Enthusiasm for Aircraft Building', *Reynold's News*, 24 March 1918. Whitehead's only SE5a order was cancelled, presumably because the company lacked the capacity to produce 100 high specification aircraft. Sturtivant and Page, *The SE5 File*, p. 11.
39. Captain Meredith Thomas and Squadron Leader (?) de Burgh quoted in Jones, *King of Air Fighters*, pp. 100–3; Lieutenant William MacLanachan quoted in Oughton and Smyth, *Ace With One Eye*, pp. 148–9.
40. 'McScotch', *Fighter Pilot*, p. 15.
41. Ibid.
42. Ibid., p. 159; Arthur Rhys Davids (like Ball, Lewis and McCudden specially selected to fly in 56 Squadron) quoted in Steel and Hart, *Tumult In The Clouds*, p. 291.
43. 'McScotch', *Fighter Pilot*, p. 50
44. Ibid., pp. 49–50.
45. Lewis, *Sagittarius Rising*, pp. 136–7. Lewis offered graphic descriptions of mess rags and the 'frenzied rush' of London leave, ibid., pp. 87–90 and 136–8.
46. In June 1917, in response to a Gotha raid on the capital, 56 Squadron was briefly brought back from France, amid much publicity. Stationed in Kent, pilots took every opportunity to motor up to London and have some fun. Later, Home Defence was expanded in a more systematic, planned – and disciplined – fashion. Ibid., pp. 183–95

47. 'McScotch', *Fighter Pilot*, p. 49.
48. Ibid., pp. 57–8.
49. 'In the course of many actions he has *driven off* a large number of enemy machines, and *has forced down three balloons*, showing a very fine offensive spirit and great fearlessness in attacking the enemy at close range and low altitudes' (author's italics), MC citation, *London Gazette*, 17 September 1917, replicated in *Historic Aircraft, Aeronautica and Medals*, Sotheby's catalogue, 19 September 1992. My thanks to Cally Sherlock of Sotheby's for her help re the award and provenance of Mannock's medals.
50. 20 July 1917, Oughton (ed.), *The Personal Diary of 'Mick' Mannock*, pp. 115–27. Mannock turned his Nieuport over in a cornfield behind Allied lines, a tappet rod having broken free and ripped the cowling off, forcing him to descend from 15 000 feet powerless. Ibid., p. 121. MacLanachan's portrait of 40 Squadron in the summer of 1917 unintentionally brought out deep tension and rivalries. His naive if well-intentioned gesture of having the MC ribbon secretly sown on Mannock's tunic may have added to the latter's pleasant surprise, but no doubt irritated Godfrey et al. 'McScotch', *Fighter Pilot*, p. 79.
51. Edward Mannock to Jim Eyles, 13 July 1917, in Jones, *King of Air Fighters*, p. 135.
52. The Eyleses later received a macabre parcel of souvenirs, including the pilot's boots and more fabric. A mutual friend from Wellingborough and another pilot who won the MC, Archie Reeves, was drinking with Mannock when he posted the parcel, and he too received a piece of fabric. Edward Mannock to Jim Eyles, 1 August 1917, ibid., pp. 143–4. Captain Reeves's middle son has always been known within his family as 'Paddy' in memory of Mannock, and his widow retains the souvenir. My thanks to Bob Reeves for this information.
53. 20 July 1917, Oughton (ed.), *The Personal Diary of 'Mick' Mannock*, pp. 117–19. 'McScotch', *Fighter Pilot*, pp. 67–9. Mannock was also affected by his having killed the observer's dog. Six months later he also acquired a terrier, albeit one that stayed on the ground.
54. Ironically, Bond himself had applauded Mannock's MC, telling his wife that, 'He is absolutely without fear and does his job always.' W.A. Bond quoted in Jones, *King of Air Fighters*, p. 139.
55. 'McScotch', *Fighter Pilot*, pp. 16–17 and 89–95.
56. Captain G.L. Lloyd quoted in Jones, *King of Air Fighters*, p. 114.
57. Edward Mannock quoted in ibid.
58. 'McScotch', *Fighter Pilot*, p. 95.
59. Ibid., pp. 97, 113 and 137
60. Ibid., p. 70–5.
61. Ibid., pp. 84–8, 90–4, 99–106 and 109; 19 August 1917, Oughton (ed.), *The Personal Diary of 'Mick' Mannock*, pp. 127–31 and 165 footnote 102; Edward Mannock to Jim Eyles, 1 August 1917, Jones, *King of Air Fighters*, pp. 143–4; Dudgeon, '*Mick*', p. 100.
62. 19 and 31 August 1917, Oughton (ed.), *The Personal Diary of 'Mick' Mannock*, p. 133.
63. 31 August 1917, ibid. p. 137. Some of the later Nieuports, including an aircraft flown by MacLanachan, carried a double Lewis machine gun, which fired quicker than forward-firing, synchronized guns and could destroy enemy aircraft from beneath. 'McScotch', *Fighter Pilot*, pp. 145–6.

64. Ibid., p. 136.
65. Ibid., pp. 110–12.
66. Ibid., pp. 133–5.
67. Ibid., pp. 148–53. 'One of my boys got another one down, and they got poor old Kennedy...we ought to have got the lot. It was really a disappointing show': 31 August 1917, Oughton (ed.), *The Personal Diary of 'Mick' Mannock*, pp. 135–7.
68. 'McScotch', *Fighter Pilot*, p. 152.
69. 31 August 1917, Oughton (ed.), *The Personal Diary of 'Mick' Mannock*, pp. 139 and 168 footnote 108.
70. 5 September 1917, ibid., pp. 139–45 and 168 footnote 109. The 'flamer' came back to haunt Mannock when months later he received a request that had been dropped over the British lines asking for details of Frech's death to be sent to his father in Konigsberg. He immediately wrote to the parents expressing his deep regret at their son's death.
71. Jones, *King of Air Fighters*, pp. 169–70; Steel and Hart, *Tumult In The Clouds*, pp. 326–7; McCudden, *Five Years in the Royal Flying Corps*, pp. 220, 223 and 309; 'McScotch', *Fighter Pilot*, pp. 66–8 and 73–4. Mannock's black humour was already evident in mess speeches with 40 Squadron, most notably his farewell address: William Douglas quoted in Dudgeon, *'Mick'*, p. 114.
72. Edward Mannock to Jim Eyles (?) November 1917, quoted in Jones, *King of Air Fighters*, p. 153. 'On one occasion he attacked a formation of five machines single-handed and shot one down out of control', extract from Bar to MC citation, *London Gazette*, 18 October 1917, replicated in *Historic Aircraft, Aeronautica and Medals*, i.e. Mannock partly received his award for action which his critics in 40 Squadron deemed reckless.
73. When a Canadian platoon wanted to shoot a downed German pilot who had fired at a balloon observer, Mannock stayed in the trenches defending his prisoner with a Very pistol until help arrived. 'McScotch', *Fighter Pilot*, pp. 156–61, 172 and 180.
74. Ibid., pp. 155 and 162. William Douglas quoted in Dudgeon, *'Mick'*, p. 112.
75. 'McScotch', *Fighter Pilot*, p. 155. Gwilym Lewis, four years McElroy's junior, also saw him as a 'wild youngster': Gwilym Lewis to Hugh and Grace Lewis, 28 June 1918, Bowyer (ed.), *Wings Over The Somme 1916–18*, p. 138.
76. 'Major George Edward Henry McElroy', WO339/110067, PRO.
77. '...in July 1918...I remember him telling McElroy then, "Don't throw yourself away, don't go down to the deck...don't do that. You'll get shot down from the ground!" And ultimately that's what happened to him and indeed to Mick Mannock too': IWM interview with Gwilym Lewis quoted in Steel and Hart, *Tumult In The Clouds*, p. 327; Gwilym Lewis to Hugh and Grace Lewis, 19 February 1918, Bowyer (ed.), *Wings Over The Somme 1916–18*, p. 116.
78. Ibid.; 'McScotch', *Fighter Pilot*, pp. 226–9.
79. William Douglas quoted in Dudgeon, *'Mick'*, p. 112.
80. Ibid., p. 113.
81. 'McScotch', *Fighter Pilot*, pp. 164–5; Gwilym Lewis to Hugh and Grace Lewis, 10 and 19 December 1917 and 7 January 1918, and Introduction, Bowyer (ed.), *Wings Over The Somme 1916–18*, pp. 99, 101, 105 and 16.
82. Barker, *The Royal Flying Corps in France*, pp. 161–7.
83. 'McScotch', *Fighter Pilot*, p. 178.

84. Ball had played a key role in modifying the SE5: 'Between the two of us we finally cooked up the SE5 so that it really worked. General Trenchard approved finally and all our other aeroplanes were altered to the same design...' IWM interview with Engineering Officer H. N. Charles, 56 Squadron, quoted in Steel and Hart, *Tumult In The Clouds*, p. 212.
85. The arrival of the SE5a coincided with the onset of winter, and eventually the rest of A flight refused to fly for long periods with Mannock in such icy temperatures. McElroy secured a compromise ceiling of 14 000 feet. McCudden graphically described the ferocious effect of severe oxygen deprivation *and* severe weather once the pilot dropped to a lower altitude. 'McScotch', *Fighter Pilot*, pp. 200 and 202–3; McCudden, *Five Years in the Royal Flying Corps*, pp. 264–5.
86. Ibid., pp. 178–9 and 186–7; Edward Mannock to Jim Eyles, 9 December 1917, Jones, *King of Air Fighters*, p. 154; Sturtivant and Page, *The S.E.5 File*, p. 6. On the operational impact of so many unreliable engines, see also Gwilym Lewis to Hugh and Grace Lewis, 30 December 1917, Bowyer (ed.), *Wings Over The Somme 1916–18*, p. 103.
87. Dudgeon, '*Mick*', p. 110; 'McScotch', *Fighter Pilot*, pp. 188–9; Major J. L. T. Pearce quoted in Jones, *King of Air Fighters*, p. 131.
88. Oughton claimed Mannock's dislike of Trenchard was rooted in the operational ban on parachutes, although it now appears the latter's rival, David Henderson, was the most vocal in opposing any suggestion that the RFC/RAF emulate the Germans in providing a means of escape from a burning aircraft. Oughton (ed.), *The Personal Diary of 'Mick' Mannock*, p. 154 footnote 38; Barker, *The Royal Flying Corps in France*, pp. 80–8.
89. Ibid., p. 199; 'McScotch', *Fighter Pilot*, p. 214; Dudgeon, '*Mick*', pp. 107–8.
90. Ibid., p. 110; 'McScotch', *Fighter Pilot*, pp. 187–8; Sturtivant and Page, *The S.E.5 File*, p. 6. Albert Ball to Albert Ball (senior), 22 March and 11 and 13 April 1918, Correspondence of Captain Albert Ball, NRO.
91. Barker, *Royal Flying Corps in France*, p. 198; 'McScotch', *Fighter Pilot*, p. 205.
92. In 56 Squadron Cecil Lewis believed from the outset that the SE5a was, 'a triumph of human intelligence and skill – almost a miracle'. Lewis, *Sagittarius Rising*, p. 182.
93. Gwilym Lewis to Hugh and Grace Lewis, 30 December 1917, Bowyer (ed.), *Wings Over The Somme 1916–18*, p. 103. Between 23 and 29 December 1917 McCudden secured nine victories, a feat which helped him secure the VC in April 1918. Cooksley, 'J. T. B. McCudden', pp. 146–7.
94. Ibid., pp. 147–9. 'McScotch', *Fighter Pilot*, pp. 210–11.
95. C.G. Grey, 'Introduction' in McCudden, *Five Years in the Royal Flying Corps*, pp. xi–xiv.
96. Major-General Sir Hugh Montague Trenchard, 'Prefatory Notes', ibid., p. ix.
97. Ibid., pp. 134–5 and 217–18.
98. Ibid., pp. 232–3. Cecil Lewis identified a keen attention to gunnery as a key element in the make-up of a 'crack fighter', as did 'Billy' Bishop; Lewis, *Sagittarius Rising*, pp. 171–2; Major William Bishop VC, *Winged Warfare* (London: Hodder & Stoughton, 1918), pp. 72–3.
99. Steel and Hart, *Tumult In The Clouds*, p. 310.
100. Lieutenant John Grider (American volunteer) quoted in ibid., p. 243: 'The rest of the squadron objected because he was once a Tommy and his father

was a sergeant-major in the old army. I couldn't see that that was anything against him but the English have great ideas of caste.' Mannock's similar service/family background would not have been common knowledge in the same way as that of the McCudden siblings.
101. Yeates's novel suggested that by early 1918 Mannock and McCudden were already familiar names in the trenches. Yeates, *Winged Victory*, pp. 52–3 and 83.
102. Mannock never saw MacLanachan again, the latter spending the rest of the war at home in an RFC backwater.
103. 'Major Robert Gregory MC', WO339/42377 and WO389/4, PRO; W.B. Yeats, 'Robert Gregory – A Note of Appreciation', *Observer*, 17 February 1917.
104. See Chapter 6.
105. 'Reprisals' was posthumously published in 1948. David Pierce, *Yeats's Worlds: Ireland, England and the Poetic Imagination* (New Haven, CT: Yale University Press, 1995) pp. 210–11.
106. Mannock and MacLanachan's ground crew stopped the latter from flying on Christmas Day in protest at what he saw as the hypocrisy of joyous celebration in the midst of carnage. 56 Squadron did not relax, and McCudden insisted on flying. McCudden, *Five Years in the Royal Flying Corps*, p. 297; 'McScotch'. *Fighter Pilot*, pp. 217–20.
107. Ibid., p 231; Gwilym Lewis to Hugh and Grace Lewis, 30 December 1917 and 7 January 1918, Bowyer (ed.), *Wings Over The Somme 1916–18*, pp. 103 and 105; Major Tilney quoted in Jones, *King of Air Fighter*, p. 158.
108. W.G. Soltau (Canadian pilot in 40 Squadron) quoted in ibid.
109. IWM interview with Gwilym Lewis quoted in Steel and Hart, *Tumult In The Clouds*, pp. 322–3.

Chapter 5   74 and 85 Squadrons, 1918

1. Jim Eyles quoted in Dudgeon. '*Mick*', pp. 117–18.
2. Ibid., pp. 118–19.
3. Edward Mannock to Jim Eyles, 4 February 1918, quoted in Jones, *King of Air Fighters*, p. 159.
4. Barker, *The Royal Flying Corps in France*, pp. 20, 50–1, 75–6, 187 and 199; Steel and Hart, *Tumult In The Clouds*, pp. 15, 24, 64, 111–12, 269–74 and 307–8; John Sweetman, 'The Smuts Report of 1917', *Journal of Strategic Studies*, IV (1981) pp. 152–74. Trenchard actually resigned as CAS on 18 March 1918, but was persuaded by Rothermere to delay his departure given the imminent launch of the RAF on 1 April and the critical situation in France. He was succeeded by Sykes on 13 April with the minister himself resigning within a fortnight. In June 1918 Trenchard was commissioned to create the British Independent Bombing Force, and the following October he assumed command of a joint Allied strategic bombing force.
5. Jim Eyles quoted in Jones, *King of Air Fighters*, pp. 159–60, and Dudgeon, '*Mick*', p. 118.
6. Edward Mannock to Mary Lewis, 6 February 1918, Edward Mannock Papers, Item 1839, IWM.

7. Edward Mannock to Jim Eyles, 11 February 1918, in Dudgeon, 'Mick', p. 118; Sturtivant, 'British Flying Training in World War I', p. 24.
8. Jones acquired a reputation between the wars as a hard drinker, but in 1918 he was a teetotaller, convinced that 'many fellows have been shot down as the result of dulled wits'. 26 April 1918, Ira Jones, *Tiger Squadron* (London: White Lion Publishers, 1972) p. 89.
9. Ira Jones, *An Air Fighter's Scrapbook* (London: Nicholson & Watson, 1938; Greenhill Books, 1990) pp. 19–26; 'The Traveller', 'Where Achilles Lies', *Saga Magazine* (April 1999) pp. 46–9.
10. Major Keith Caldwell, 74 Squadron's second CO, also kept a diary, made available to Jones and Dudgeon.
11. Jones, *King of Air Fighters*, p. 167. In addition to Book IV of *King of Air Fighters* and a chapter in *An Air Fighter's Scrapbook*, see diary entries in *Tiger Squadron*.
12. Ibid., p. 62.
13. Jones, *King of Air Fighters*, p. 160.
14. Ibid., pp. xi and 160–5.
15. Ibid., pp. 163–4.
16. Ibid., pp. 160–1, 164–5 and 190; Ira Jones to Vernon Smyth, 1 May 1957, Edward Mannock papers, Misc. 117 Item 1839, IWM; Major (later Air Commodore) Keith L. Caldwell quoted in Dudgeon, 'Mick', pp. 124–5; ibid., pp. 119–20 and 122.
17. A.S.W. Dore was 35 in 1918, and in four years of war had seen service first with the Worcestershire Regiment and then as a flight commander under Sholto Douglas in 43 Squadron. Like his squadron commander he enjoyed a successful career in the interwar RAF. His diary is heavily drawn upon in the second volume of Ralph Barker's *The Royal Flying Corps in France*.
18. Jones, *King of Air Fighters*, pp. 166–7.
19. Ibid., p. 168; Jones, *Tiger Squadron*, pp. 64–5.
20. Major Keith L. Caldwell quoted in Dudgeon, 'Mick', pp. 123.
21. Lieutenant Harris G. Clements quoted in ibid., p. 124.
22. Ibid., p. 125.
23. Jones, *Tiger Squadron*, pp. 66–7; 11 April 1918, ibid., p. 72.
24. 'They [40 Squadron] gave me a good time. They've had some bad luck lately, but seem to be emerging from the despondency somewhat.' Edward Mannock to Mary Lewis, 11 April 1918, Edward Mannock Papers, Item 1839, IWM.
25. Tim Travers, *The Killing Ground: the British Army, the Western Front and the Emergence of Modern Warfare 1900–1918* (London: Allen & Unwin, 1987) pp. 220–2; Barker, *The Royal Flying Corps in France*, pp. 220–1 and 238–40; Steel and Hart, *Tumult In The Clouds*, pp. 315–18.
26. Field Marshal Sir Douglas Haig, C-in-C BEF, order of the day, 12 April 1918; Jones, *An Air Fighter's Scrapbook*, p. 74.
27. Lieutenant Harris G. Clements quoted in Dudgeon, 'Mick', p. 126; 12 April 1918, Jones, *Tiger Squadron*, p. 79.
28. Lieutenant Harris G. Clements quoted in Dudgeon, 'Mick', pp.129–30.
29. Major Keith L. Caldwell quoted in ibid., pp. 127.
30. 30 April 1918, Jones, *Tiger Squadron*, p. 91; Lieutenant (?) Dolan letter home quoted in Dudgeon, 'Mick', p. 129.
31. 19 April and 18 May 1918, Jones, *Tiger Squadron*, pp. 82 and 111; Jones, *King of Air Fighters*, p. 231.

32. With a New Zealand captain, and an England fly-half (Harry Coverdale) – as well as Jones who later played scrum-half for the RAF – 74 won all of its 18 inter-squadron matches. Mannock's ferocious tackling more than made up for an absence of ball-handling skills. Ibid., p. 185.
33. Description of mess life based on: K.L. Caldwell to Vernon Smyth, 30 May 1957, Misc. 117 Item 1839, Edward Mannock papers, IWM; Jones, *King of Air Fighters*, pp. 207–10. Jones suggested Caldwell was more prone to practical jokes than he later wished to admit.
34. Jones, *An Air Fighter's Scrapbook*, pp. 77–8. Caldwell had of course previously been awarded the MC.
35. Lieutenant W.B. Giles quoted in Dudgeon, '*Mick*', p. 128.
36. 30 April 1918, Jones, *Tiger Squadron*, p. 93.
37. Edward Mannock to Jim Eyles, 13 April 1918, in Jones, *King of Air Fighters*, p. 183.
38. Ibid., p. 188; 21 April 1918, Jones, *Tiger Squadron*, pp. 84–5.
39. Gwilym Lewis to Hugh and Grace Lewis, 24 April 1918, Bowyer (ed.), *Wings Over The Somme 1916–18*, p. 130.
40. 12 May 1918, Jones, *Tiger Squadron*, p. 102, and *King of Air Fighters*, pp. 198–9.
41. 13 and 17 May 1918, Jones, *Tiger Squadron*, pp. 104–5. Dudgeon mistakenly conflates this incident with another occasion when Mannock accused a new pilot of cowardice for avoiding a fight, but Sifton was not in A flight and had yet to see action. Dudgeon, '*Mick*', pp. 131–2.
42. Lieutenant Harris G. Clements quoted in ibid., pp.135–6.
43. Ibid., p. 135. Confirmed by Ira Jones: 'If they assume a slightly sickly appearance when they smile, he's suspicious. One of these fellows nearly vomited.' 17 May 1918, Jones, *Tiger Squadron*, p. 109.
44. Ibid., p. 108; Jones, *King of Air Fighters*, pp. 214, and 170–1. In mid-May Mannock was forced to replace his regular aircraft (C1112) with another Royal Aircraft Factory Viper-engined SE5a (C1126). In total, November 1917 – July 1918, he flew seven different aircraft. Sturtivant and Page, *The SE5 File*, pp. 53 and 140. For details of which SE5a was responsible for which victories, see ibid., p. 53.
45. Lieutenant Harris G. Clements quoted in Jones, *King of Air Fighters*, p. 211.
46. Ibid., p. 223; J.F. Hunt quoted in ibid., p. 210.
47. IWM interview with Norman Macmillan, quoted in Steel and Hart, *Tumult In The Clouds*, pp. 247–8.
48. Barker, *The Royal Flying Corps*, pp. 63, 172 and 197; Jones, *King of Air Fighters*, p. 230; K.L. Caldwell to Vernon Smyth, 30 May 1957, Misc. 117 Item 1839, Edward Mannock papers, IWM. On the future CAS's First World War career, see Lord Douglas of Kirtleside, *Years of Combat* (London: Collins, 1963).
49. Barker, *The Royal Flying Corps in France*, p. 87.
50. 17 May 1918, Jones, *Tiger Squadron*, p. 108; Edward Mannock to Jim Eyles, 29 May 1918, Jones, *King of Air Fighters*, pp. 214–15. 74 beat 56 Squadron's record by ten days for destroying 50 enemy aircraft (and on 18 June did the same when reaching 100): ibid.
51. 21 and 22 May 1918, Jones, *Tiger Squadron*, pp. 119–21.
52. K.L. Caldwell to Vernon Smyth, 30 May 1957, Misc. 117 Item 1839, Edward Mannock papers, IWM.

53. DSO and Bar to DSO citations, *London Gazette*, 16 September 1918, replicated in *Historic Aircraft, Aeronautica and Medals*.
54. 24 May 1918, Jones, *Tiger Squadron*, p. 123. Caldwell confirmed Mannock's over-casual and even slovenly dress, '& this had to be watched at times when V.I.P.s paid visits'; K.L. Caldwell to Vernon Smyth, 30 May 1957, Misc. 117 Item 1839, Edward Mannock papers, IWM.
55. Gwilym Lewis to Hugh and Grace Lewis, 24 May 1918, Bowyer (ed.), *Wings Over The Somme 1916–18*, pp. 133–4.
56. 24 May 1918, Jones, *Tiger Squadron*, p. 124; A.C. Hamer, *A Short History of No. 85 Squadron* (Northampton: Services Publishing Services, 1987) pp. 5–7.
57. 29 May and 1 June 1918, Jones, *Tiger Squadron*, pp. 129–30 and 135–6.
58. Edward Mannock to Jim Eyles, 25 May and 7 June 1918, Jones, *King of Air Fighters*, pp. 207 and 228.
59. 'I think my guns must have gone wrong (not properly aligned) as I couldn't miss the damned Hun.' Edward Mannock to Major S.E. Parker, 7 June 1918, B3167, Aviation Records Department, RAFM.
60. Gwilym Lewis to Hugh and Grace Lewis, 6 June 1918, Bowyer (ed.), *Wings Over The Somme 1916–18*, p. 137.
61. Mary Lewis would have been seen at this time as a girl with spirit. She claimed to be the first woman to fly in an SE5, her brother taking her up when temporarily in command of the CFS squadron at the end of the war. Gwilym Lewis to Hugh and Grace Lewis, 20 November 1918, ibid., p. 145.
62. Edgar McLaughlin to Mary Lewis, 12 June 1918, Edward Mannock Papers, Item 1839, IWM.
63. Edward Mannock to Mary Lewis, 7 June and 10 July 1918, ibid.
64. For a Camel pilot's view of German tactics in the spring of 1918, see Yeates, *Winged Victory*, p. 146.
65. Ibid., pp. 227–8; James Dudgeon, '"Mick" Fact, Fiction and Legend: an examination of Major Edward Mannock VC DSO MC', *Cross and Cockade* (1980) XI, p. 109; 16 June 1918, Jones, *Tiger Squadron*, p. 137.
66. 17 June 1918, ibid., pp. 138–9.
67. 18 June 1918, ibid., p. 140.
68. Major Keith L. Caldwell quoted in Dudgeon, '*Mick*', p. 155.
69. Edward Mannock to Jess Mannock, 16 June 1918, Jones, *King of Air Fighters*, p. 232.
70. Jones, *An Air Fighter Scrapbook*, pp. 68–9.
71. Ibid., p. 69; Dudgeon, '*Mick*', pp. 140–7.
72. Jim Eyles quoted in ibid., p. 154.
73. Ibid., pp. 154–5; Major Keith L. Caldwell quoted in ibid., p. 155; Edward Mannock to Jim Eyles, 18 June 1918, Jones, *King of Air Fighters*, p. 235; Hamer, *A Short History of No. 85 Squadron*, p. 7. Also waiting at the RAF Club was confirmation of a second bar to Mannock's DSO, awarded on account of his 48 victories, 'due to wonderful shooting and a determination to get to close quarters; to attain this he displays most skilful leadership and unfailing courage'. Second Bar to DSO citation, *London Gazette*, 16 September 1918, replicated in *Historic Aircraft, Aeronautica and Medals*
74. Jim Eyles quoted in Dudgeon, '*Mick*', p. 154.
75. Jim Eyles quoted in Dudgeon, '*Mick*', pp. 154–5; Dorothy Mannock quoted in ibid., p. 155.

76. Lord Southborough's committee, mostly composed of army officers (even the doctors), sat from September 1920 to June 1922; see Ted Bogacy, 'War Neurosis and Cultural Change in England 1914–1922: the work of the War Office Committee of Enquiry Into "Shell-Shock"', *Journal of Contemporary History*, XXIV (1989) pp. 227–56.
77. Samuel Hynes, *A War Imagined: The First World War and English Culture* (London: Bodley Head, 1990), p. 307. Hynes noted that in 1920 there were 65 000 ex-soldiers receiving disability pensions for neurasthenia, of which 9000 were still hospitalized.
78. Evidence of Captain W. H. R. Rivers MD in Army, *War Office Committee of Enquiry Into 'Shell-Shock'* (London: HMSO, 1922) Cd. 1734, pp. 55–8; W. H. R. Rivers, 'Fear and its expression', *Instinct and the Unconscious* (Cambridge: Cambridge University Press, 1922), p. 209; Eric J. Leed, *No Man's Land: Combat and Identity in World War I* (Cambridge: Cambridge University Press, 1979) pp. 180–1. Leed noted Graves's remark that 'every man who spent more than three months under fire could legitimately be considered neurasthenic', and John Keegan drew on British Army research 1940–5 and the US Army's 1946 report on *Combat Exhaustion* to suggest that even in a far less static war an infantryman would break down just inside a year. Robert Graves quoted in ibid., p. 181; Keegan, *The Face of Battle*, pp. 328–9.
79. Evidence of Squadron Leader E.W. Craig in Army, *War Office Committee of Enquiry Into 'Shell-Shock'*, pp. 86–7.
80. W.M. Maxwell quoted in Leed, *No Man's Land*, p. 167; Edward Mannock to Mary Lewis, 6 February 1918, Edward Mannock Papers, Item 1839, IWM; Barker, *The Royal Flying Corps in France*, pp. 85–8.
81. F.C. Bartlett, *Psychology and the Soldier* (Cambridge: Cambridge University Press, 1927) pp. 192–3. Bartlett drew heavily upon wartime research summarized in: J.T. MacCurley, *War Neuroses* (Cambridge: Cambridge University Press, 1918).
82. Siegfried Sassoon quoted in Leed, *No Man's Land*, p. 22. Pat Barker's fictional depiction of W. H. R. Rivers points out to Sassoon that, 'Taking *unnecessary* risks is one of the first signs of a war neurosis.' Pat Barker, *Regeneration* (London: Penguin, 1992) p. 12.
83. Bartlett, *Psychology and the Soldier*, p. 193; Edward Mannock quoted by Jim Eyles in Dudgeon, '*Mick*', p. 154.
84. Gwilym Lewis to Hugh and Grace Lewis, 18 July 1918, Bowyer (ed.), *Wings Over The Somme 1916–18*, p. 142.
85. Gwilym Lewis to Hugh and Grace Lewis, 21 July 1918, ibid. Lewis was probably designated 'Flying Sickness D. [debility]', an all-embracing category which facilitated sick leave and a home posting.
86. A second phase, where the virus was clearly more virulent and potentially fatal, reached London in August. Pete Davies, *Catching Cold: 1918's forgotten tragedy and the scientific hunt for the virus that caused it* (London: Michael Joseph, 1999) pp. 59–60. Gwilym Lewis to Hugh and Grace Lewis, 28 June 1918, ibid. The editor must have transcribed the date incorrectly from the original letter as it refers to meeting 'old Mannock on his return to France'.
87. Jones, *King of Air Fighters*, pp. 235–6.
88. Elliot White Springs, Larry Callaghan, and John McGavock Grider, were known as the 'Musketeers'. Springs's daily impressions of war on the Western

Front were attributed to the dead Grider in the anonymously authored *War Birds – Diary of an Unknown Aviator* (London: Hamilton, 1926).
89. Lieutenant C.B.R. MacDonald quoted in Jones, *King of Air Fighters*, pp. 236–7; Dudgeon, *'Mick'*, pp. 157–8.
90. Jones, *King of Air Fighters*, pp. 237–8.
91. Cooksley, 'J.T.B. McCudden', p. 149.
92. Jones, *King of Air Fighters*, pp. 240–1.
93. Ibid., p. 241; Dudgeon, *'Mick'*, pp. 159–60.
94. Steel and Hart, *Tumult In The Clouds*, pp. 330 and 334; Ferguson, *The Pity of War*, pp. 310–11.
95. Gwilym Lewis to Hugh and Grace Lewis, 21 July 1918, Bowyer (ed.), *Wings Over The Somme 1916–18*, p. 142.
96. Photographs of Major Mannock in Hamer, *A Short History of No. 85 Squadron*, p. 8, and of 85 Squadron on 25 July 1918 in Dudgeon, *'Mick'*, centre section. Nurses were under acute pressure in July 1918 as the bulk of hospital provision in St Omer had been set aside for influenza victims, most of whom needed constant fluids, but were also incontinent. Lyn Joseph *The Roses of No-Man's Land* (London: Michael Joseph, 1980; Papermac, 1984) pp. 285–7.
97. Jones, *King of Air Fighters*, p. 245; Oughton and Smyth, *Ace With One Eye*, pp. 278–9; Dudgeon, *'Mick'*, p. 161.
98. Jones, *King of Air Fighters*, pp. 245–6.
99. Ibid., p. 246.
100. Lieutenant Donald Inglis quoted in Jones, *King of Air Fighters*, p. 248.
101. This account of Mannock's death is based upon ibid., pp. 248–9; Edward Naulls (private, D company, 2nd Battalion, Essex Regiment, entrenched between Robecq and Pacaut Wood, 26 July 1918) to Reginald Pound (editor, *Strand*), 1 November 1943, AC 72/10 13, Edward Mannock papers, Aviation Records Department, RAFM; Lieutenant D.C. Inglis, combat report, 26 July 1918, Misc. 117 Item 1839, Edward Mannock papers, IWM; Dudgeon, *'Mick'*, pp. 19–22, and '"Mick" Fact, Fiction and Legend', pp. 97 and 105–9.
102. Lieutenant Donald Inglis quoted in Jones, *King of Air Fighters*, p. 249. Different contemporary accounts have Inglis rescued by a company either of the 24th Welch or of the Suffolk Regiment.
103. Ibid., pp. 249–51.
104. Lieutenant W.E.W. Cushing to Sergeant Patrick Manning, 26 July 1918, AC 72/10 21, Edward Mannock papers, Aviation Records Department, RAFM.
105. Rev. Bernard Keymer, CF, to Sergeant Patrick Mannock, 28 July 1918, AC 72/10 22, ibid.; Lieutenant Malcolm McGregor, diary, quoted in Hart and Steel, *Tumult In The Clouds*, p. 329.
106. Ira Jones's diary, 27 July 1918, quoted in ibid., pp. 251–2.
107. Correspondence surrounding death of Major Edward Mannock, 1918–21, WO339/66665/142133, PRO.
108. 'A Great Airman. Major Mannock's Leadership', *The Times*, 9 August 1918.
109. Correspondence surrounding death of Major Edward Mannock, 1918–21, WO339/66665/142133, PRO.
110. Patrick Mannock eventually disposed of all of Edward's effects other than his identity discs. Dudgeon, *'Mick'*, pp. 168 and 176.
111. IWGC to Rev. E. Rogers (Church Lads' Brigade), 2 June 1921, in Jones, *King of Air Fighters*, p. 254; IWGC to Jim Eyles (date not known), ibid., pp. 254–6.

112. Dudgeon, 'Mick', pp. 172–6.
113. David Rose, 'Last resting place of lost wartime air ace?', *Kentish Gazette*, 5 November 1993.
114. Bourke, *Dismembering the Male*, pp. 228–9 and 235–6.

Chapter 6 Conclusion

1. From Ivor Gurney, 'Laventie', in P.J. Kavanagh (ed.), *Collected Poems of Ivor Gurney* (Oxford: Oxford University Press, 1982) pp. 77–8. Pte Gurney (2nd/5th Gloucester Regiment) was in the trenches at Laventie, June–July 1916.
2. Grave 12, Row F, Plot 3, Commonwealth War Graves Commission Military Cemetery, Laventie, west of Lille.
3. Angus Stewart and Keith Caldwell quoted in Dudgeon, ' "Mick" Fact, Fiction and Legend', pp. 104–5.
4. The family has suggested that, if the body was Mannock, then it was found intact because he jumped out of the burning aircraft as it glided down to earth.
5. Lieutenant Harris G. Clements quoted in Dudgeon, *Mick*, p. 180.
6. Major Keith L. Caldwell quoted in ibid., pp. 179–80; K.L. Caldwell to Vernon Smyth, 30 May 1957, Misc. 117 Item 1839, Edward Mannock papers, IWM.
7. Jones, *King of Air Fighters*, pp. 259–60; Oughton and Smyth, *Ace With One Eye*, pp. 295–6; 'VC for Late Major Mannock', *Kentish Gazette and Canterbury Press*, 26 July 1919.
8. 'Ronald McNeil', *Who's Who 1918* (London: A. & C. Black, 1918) p. 1564.
9. 'Proposals by Mr Churchill for posthumous awards', 1919–21, AIR2/106/A17376, PRO.
10. Yet proportionately the number of genuine aces in the two world wars were not that different: consistent with aerial combat's principle of rigorous natural selection based upon what jet pilots label 'acute situational awareness', 5 per cent of RFC/RAF fighter pilots accounted for over half of all 'kills', and in 1939–45 5 per cent of Fighter Command accounted for 40 per cent of all victories. My thanks to Commander P.J. Johnston for pointing these figures out to me.
11. Group Captain C.L.N. Newall to Colonel M.D. Graham, (?) 1920, AIR2/91/C51021, PRO. Mannock etc. are simply listed as 'France (Flying Services)' as opposed to the specific date and place where an individual act of heroism took place. War Office, *Alphabetical List of Recipients of the Victoria Cross, during the War, 1914–20* (London: HMSO, 1920).
12. Draft citation for award of VC to Capt. (Actg Maj.) E. Mannock, July 1919, AIR2/91/C90866, PRO; VC citation, *London Gazette*, 18 July 1919, replicated in *Historic Aircraft, Aeronautica and Medals*.
13. Steel and Hart, *Tumult In The Clouds*, pp. 167–8 and 240; T.E. Shaw (Lawrence) to Ira Jones, (?) 1935, in Jones, *An Air Fighter's Scrapbook*, pp. 198–9.
14. Leed, *No Man's Land*, pp. 134–5.
15. Michael Paris, 'The Rise of the Airmen: the Origins of Air Force Elitism c. 1890–1918', *Journal of Contemporary History*, XXVIII (1993) pp. 124–5.
16. Ibid., pp. 134–7; John Buchan, *Mr Standfast* (London: Thomas Nelson, 1919; Penguin, 1997) p. 19; Michael Paris, *From the Wright Brothers to Top Gun:*

Aviation, Nationalism and Popular Cinema (Manchester: Manchester University Press, 1995) p. 28; Barker, *The Royal Flying Corps in France*, pp. 96–7.
17. Edward Mannock to Major S.E. Parker, 7 June 1918, B3167, Aviation Records Department, RAFM.
18. AAF to Air Ministry (telegram), and notes of Under-Secretary of State, 3 July 1919, AIR2/91/C90866, PRO.
19. Full list of Mannock's confirmed and unconfirmed victories in Dudgeon, '"Mick" Fact, Fiction and Legend', pp. 108–9.
20. In 1938 Jones went even further, absurdly claiming that Mannock shot down over 100 aircraft. Jones, *An Air Fighter's Scrapbook*, pp. 69.
21. For a summary of the case for and against Bishop see Barker, *The Royal Corps in France*, pp. 96–7 and 243. Even a commissioned history of 85 is implicitly critical: Hamer, *A Short History of 85 Squadron*, p. 8.
22. On 85 Squadron's reputation for rowdiness when Bishop was CO, see Steel and Hart, *Tumult In The Clouds*, pp. 298–9.
23. Canadian Air Force Office of Public Affairs, 'The Flying Career of William Avery Bishop', http://raven.cc.ukans.edu/~kansite/ww_one/comment/bishop.html; Jones, *An Air Fighter's Scrapbook*, p. 66.
24. Jones, *King of Air Fighters*, pp. 260–1; 'McScotch', *Fighter Pilot*, pp. 240–1 and 248.
25. Edward Mannock senior was excluded from his son's will, despite his early interest in its content (see Chapter 5). Cuthbert Gardner and Patrick Mannock were instructed to distribute the £600 legacy to all other members of the immediate family. Various possessions left in Wellingborough were retained by the Eyleses, and a pair of riding boots, a prismatic compass, and a Sam Browne remained in the loft when the family moved down the road in 1924. They remained in 2000 in the possession of the then occupants of 183 Mill Road, Mr and Mrs Underwood, and my thanks go to them for the opportunity to examine such vivid reminders of a life that had ended over 80 years ago. Oughton and Smyth, *Ace With One Eye*, p. 299.
26. Dudgeon, *'Mick'*, p. 171; information from Anne Tooke; 'The legendary V.C., D.S.O. and two bars, M.C. and bar group to Major E. "Mick" Mannock, Royal Flying Corps', *Historic Aircraft, Aeronautica and Medals*, Sotheby's, Sussex, 19 September 1992.
27. Photo-postcard sent by Edward to Patrick Mannock, 1912: 'Hope you're well. Note Demosthenes "holding forth". A lesson in socialism at our local Y.M.C.A. camp at W'bor...E.' Replicated in *Historic Aircraft, Aeronautica and Medals*, Sotheby's, Sussex, 24 November 1994. Again, thanks to Cally Sherlock of Sotheby's for information re both sales.
28. Colin Wilkinson, 'Flying ace who took one risk too many', *Kettering Evening Telegraph*, 14 September 1992.
29. Tony Smith, 'Flying ace who knew no fear', *Kettering Evening Telegraph*, 18 February 1991.
30. On the wartime origins of rolls of honour, see King, *Memorials of the Great War in Britain*, p. 45.
31. Oughton and Smyth, *Ace With One Eye*, pp. 299–300.
32. The picture is clearly posthumous as Mannock's tunic bears the VC medal ribbon. It may be the same portrait that once hung in the RAF Club, and if so it was rescued by Archie Reeves in 1923 after sustaining damage when the

club moved to Piccadilly. Major R. Ludlow to Captain A.C. Reeves, 30 January 1923, Reeves family papers.
33. To be fair the Civic Society is a wholly voluntary organization, and a new permanent exhibition is planned for larger premises. My thanks to John Goddard, Honorary Curator of the Wellingborough Heritage Centre Trust, for his advice and information, not least inspiring an abortive trip to find the Eyles family plot in London Road cemetery.
34. 'V.C. for late Major Mannock – Greatest British Airman – Suggested memorial at Canterbury', *Kentish Gazette and Canterbury Press*, 26 July 1919.
35. King, *Memorials of the Great War in Britain*, p. 129.
36. Major M.S. Marsden, 'The Amazing Mannock, V.C. No. 5 of Knights of the Air', *Sunday Graphic and Sunday News*, 30 September 1934.
37. Dyer, *The Missing of the Somme*, pp. 10–11 and 30–1.
38. On hospitals as functional memorials, see King, *Memorials of the Great War in Britain*, pp. 68, 74 and 77.
39. R. Andrew to Vernon Smyth, 2 April 1957, and G.R. Hews to Vernon Smyth, 3 April 1957, Edward Mannock papers, Misc. 117 Item 1839, IWM.
40. 'Hero who ruled the skies', *Kentish Gazette*, 21 July 1978.
41. 'No fly-past for First World War air hero', ibid., 2 August 1968. In November 1968 the RAF marked the Armistice's 50th anniversary by naming 14 VC10 transports after air aces of both world wars, including Mannock.
42. 'Mannock memorial service' and 'Nieces' own salute to war hero', ibid., 31 July 1964 and 1 August 1980. Patrick Mannock died in Edinburgh in August 1962.
43. Arguably the largest corporate gesture is in Farnborough where a replica of Mannock's SE5a hangs in the main shopping mall to remind shoppers of the town's historic links with aviation.
44. Dyer, *The Missing of the Somme*, p. 110.
45. King, *Memorials of the Great War in Britain*, p. 204.
46. 'McScotch', *Fighter Pilot*, pp. 246–8; Jones, *King of Air Fighters*, pp. 260–85.
47. Group Captain Sir Douglas Bader, 'Forward' in Dudgeon, '*Mick*', p. 15, and 'Foreword' in Air Vice-Marshal 'Johnnie' Johnson, *Wing Leader* (London: 1956; Goodall, 1995) p. 9; Paul Brickhill, *Reach For The Sky* (London: Williams Collins, 1956) p. 198.
48. Johnson, *Wing Leader*, pp. 54 and 58.
49. On Yeates's popularity with the RAF but not the critics, see Henry Williamson, 'Preface to the new edition' in Yeates, *Winged Victory*, p. 8, and Samuel Hynes, *The Soldiers' Tale: Bearing Witness to Modern War* (London: Pimlico, 1998) p. 125. On the relationship between Yeates and Henry Williamson, and the latter's role in publication of *Winged Victory*, see Anne Williamson, *Henry Williamson: Tarka and the Last Romantic* (Stroud: Alan Sutton, 1995) pp. 25, 162–3, 166–7, 172–9 and 198.
50. Squadron Leader Ronald Adams to Patrick Mannock (?) 1942, AC 72/10 Item 15, B3167, Aviation Records Department, RAFM. 74 had been reformed in September 1935, as in common with 40 and 85 it had been wound up after 1919.
51. D. Stokes, *Paddy Finucane: Fighter Ace* (London: William Kimber, 1983; Crécy Books, 1992) pp. 174–87 and 15.

52. Hillary claimed similar assumptions about the Germans' 'mass psychology' were still being passed on to new pilots in 1940: Hillary, *The Last Enemy*, p. 57. 'McScotch', *Fighter Pilot*, pp. 247 and 246; Jones, *King of Air Fighters*, p. 261. Buchan, *Mr Standfast*, pp. 145 and 307. Buchan's novel is well researched, witness the resemblance of Roylance's powerful but unreliable 'Shark-Gladas' scout to an SE5. Confirmation of how, beyond the Western Front, Mannock's fame was posthumous, is his omission from Buchan's 1918 list of 'heroes who had won their spurs since the Somme'; ibid.
53. 'McScotch', *Fighter Pilot*, pp. 246–7; Jones, *King of Air Fighters*, p. 261, and *An Air Fighter's Scrapbook*, pp. 27–8, 50–4, and 263–70.
54. 'McScotch', *Fighter Pilot*, pp. 246–7; Jones, *King of Air Fighters*, p. 261.
55. R.A.C. Parker, *Chamberlain and Appeasement: British Policy and the Coming of the Second World War* (London: Macmillan, 1993) pp. 18 and 42.
56. Jones finally resigned in 1936 to work as a freelance writer and pilot. After service in Russia in 1919 he was publicly critical of the White Russians, and complimentary of Trotsky as a strategist. Further unguarded comments after service in Iraq, plus his maverick behaviour, led to a veto on his applying to Staff College, after which his chances of promotion were minimal. Although appointed wing commander after being recalled as an instructor in 1939, heavy drinking resulted in early demobilization. Jones, *An Air Fighter's Scrapbook*, pp. 107–57, 210 and 214–15; 'The Traveller', 'Where Achilles Lies', p. 49.
57. 'McScotch', *Fighter Pilot*, pp. 247–8.
58. 'Preface', Jones, *King of Air Fighters*, p. viii. Keen, Mannock's flight commander in 40 Squadron, made a similar point when interviewed in a British Legion-sponsored documentary on the Great War: the 'warrior aces who made the Royal Flying Corps, Mannock, McCudden, Bishop and Ball' should never be forgotten, and for their sakes, although the maintenance of peace was paramount, it could not be at any price. *Forgotten Men* (Director: Sir John Hammerton, 1934).
59. Cecil, *Sagittarius Rising*, pp. 121–4, 184–6 and 210. At the time the ex-RFC creator of Biggles had more in common with Jones; see Captain W.E. Johns, 'The Mannock Spirit', *Popular Flying*, June 1935. My thanks to Anne Tooke for this reference.
60. Ibid., p. 186.
61. Jones, *An Air Fighter's Scrapbook*, pp. 328–9.
62. Ibid., pp. 220–9 and 271–6. Jones wrote for the *News of the World*, the *Sunday Pictorial* and the *Sunday Dispatch*, and was syndicated in similar newspapers across the Empire.
63. Hynes, *The Soldiers' Tale*, pp. 76–7; Leed, *Combat and Identity in World War I*, p. 135.
64. Sebastian Faulks, *The Fatal Englishman: Three Short Lives* (London: Hutchinson, 1996; Vintage, 1997) pp. 198–9; Samuel Hynes, *The Auden Generation: Literature and Politics in England in the 1930s* (London: Bodley Head, 1976) pp. 92–3.
65. Jones, *An Air Fighter's Scrapbook*, pp. 195–200; Hynes, *The Auden Generation*, pp. 92–3, 190–1 and 238–9; Brian Holden Reid, 'T.E. Lawrence and his Biographers', in Bond (ed.), *The First World War and Military History*, p. 232.
66. Appropriately, the first official historian was the literary critic Sir Walter Raleigh, whose old world charm and Cambridge chair qualified him to write

a Whig interpretation of the RFC's early combat experience. His successor, the Air Ministry's H.A. Jones, was similarly glowing in his praise of the heroic and virtuous pilots *and* their visionary commanders. Paris, 'The Rise of the Airmen', pp. 137–8.
67. Paris, *From the Wright Brothers to Top Gun*, pp. 48–9.
68. Hynes, *The Soldiers' Tale*, pp. 92–3.
69. On the metaphors of war in St-Exupéry's interwar work, notably *Night Flight* (1931), see Leed, *Combat and Identity in World War I*, p. 135.
70. 2/33 Squadron diary, quoted in William Rees, 'Introduction' to Antoine de Saint-Exupéry, *Flight to Arras* [*Pilote de guerre*, 1942] (London: Penguin, 1995) p. xviii.
71. Ibid., pp. 73, 74–5 and 36. Mannock would similarly have approved: 'Crews are being sacrificed like glasses of water hurled at a forest fire', and 'War is not an adventure. War is a disease. Like typhus'. Ibid., pp. 4 and 38.
72. Faulks, *The Fatal Englishman*, pp. 165–6; Arthur Koestler, 'In Memory of Richard Hillary', *The Yogi and the Commissar* (London: Jonathan Cape, 1945; paperback edn 1964) pp. 55–6; Hillary, *The Last Enemy*, p. 43. Koestler's essay, based on access to some of Hillary's letters, first appeared in *Horizon*, April 1943, entitled 'The Birth of a Myth'.
73. Hillary, *The Last Enemy*, p. 85 and 14–15.
74. Richard Hillary quoted in Faulks, *The Fatal Englishman*, p. 172.
75. Hillary, *The Last Enemy*, p. 97.
76. Ibid., pp. 40–1 and 44.
77. Ibid., p. 123.
78. George Orwell, 'The English Revolution' in *The Lion and the Unicorn: Socialism and the English Genius* (London: Secker & Warburg, 1941; Penguin, 1988) pp. 112–13.
79. Ibid., p. 24. Calder interpreted *The Last Enemy* as an unqualified vindication of 'the English ruling class and its Oxbridge "incubators"', and a discrediting of pre-war left-wing pacifists, but Hillary warrants a more sympathetic reading. Calder, *Myth of the Blitz*, p. 158.
80. Oughton (ed.) 'Introduction', *The Personal Diary of 'Mick' Mannock*, p. 11; Oughton and Smyth, *Ace With One Eye*, p. 13.
81. Vernon Smyth to the Editor, *Daily Telegraph*, 2 April 1957; replies to Vernon Smyth's appeal for information; Ira Jones to Vernon Smyth, 14 April 1957. Edward Mannock papers, Misc. 117 Item 1839, IWM.
82. V. Smyth, 'The true story of Major E. Mannock VC DSO MC – the groundwork of a film story: Micky Mannock VC', and related correspondence, ibid.
83. Oughton and Smyth, *Ace With One Eye*, p. 15.
84. Peter Roberts, *The Action of the Tiger*, BBC Radio 4 (Director: Nigel Bryant, 1989).
85. *Aces High* (Director: Jack Gold, 1976).
86. Robinson, *Goshawk Squadron*, pp. 70–1, 81, 172 and 143. Woolley shares Mannock's supposed penchant for nurses, and dies when he finally loses the motivation to keep going and starts fantasizing about his mistress from the local hospital.
87. Peter Townsend, 'Preface', ibid., pp. i-ii.

88. www.crossandcockade.com/main.htm. The quarterly magazine *Cross and Cockade* is of an exceptionally high standard for supposedly 'amateur historians'. Its contributors are anything but 'anoraks'.
89. Rev. Bernard Keymer quoted in Jones, *King of Air Fighters*, p. 98
90. Jim Eyles quoted in ibid., pp. 9 and xii.
91. Ibid., pp. 170–1 and xii; Dudgeon, *'Mick'*, p. 65.
92. 'McScotch', *Fighter Pilot*, p. 109. Mannock's mother and sisters seemingly refused to believe that he was capable of 'a hatred for anyone which was so bitter and unmerciful'; Jones, *King of Air Fighters*, p. 193.
93. William MacLanachan, undated recollection (1957?), Edward Mannock Papers, Misc. 117 Item 1839, IWM.
94. Keith L. Caldwell to Vernon Smyth, 30 May 1957, ibid.
95. Edward Mannock quoted in Jones, *King of Air Fighters*, p. 9.
96. Jim Eyles quoted in Dudgeon, *'Mick'*, p. 155.
97. Edward Mannock to Mary Lewis, 7 June 1918, Edward Mannock Papers, Item 1839, IWM.
98. See Martin Petter, ' "Temporary gentlemen" in the aftermath of the Great War: rank, status and the ex-officer problem', *Historical Journal*, XXXVII (1994) pp. 127–52.
99. Dudgeon, *'Mick'*, pp. 178–9.
100. Jones, *King of Air Fighters*, p. 208.
101. Andrew Thorpe, *A History of the Labour Party* (London: Macmillan, 1997) pp. 66–7; Dowse, *Left in the Centre*, pp. 130–46.
102. Ira Jones to Vernon Smyth, 1 May 1957, Edward Mannock papers, Misc. 117 Item 1839, IWM.
103. On Williamson, Lawrence, and the BUF, see Williamson, *Henry Williamson: The Last Romantic*, pp. 187–91, 220–2 and 343–4 footnotes 42 and 43. The 'Airman' in *The Orators* (1932) has been seen as a deliberately fascist creation: Charles Osborne, *W.H. Auden: The Life of a Poet* (London: Eyre Methuen, 1980) pp. 92–3.
104. Yeates, *Winged Victory*, pp. 41, 56–60, 181, and 150. Tom Cundall (Yeates) could see that 'socialism was natural to man, and individualism a disease' and believed that Jewish international financiers were responsible for the war.
105. On Wintringham's involvement in 1918–19 army protests, see T.H. Wintringham, 'Mutiny', *Left Review*, August 1935, pp. 441–6.
106. Wintringham commanded the IB's British Battalion in early 1937. Tom Wintringham, *English Captain* (London: Faber & Faber, 1939) pp. 37–8; Brian Bridgeman, *The Flyers* (Swindon: Brian Bridgeman, 1989) pp. 103, 161, 164–5 and 167–9. The nearest the British pilots had to a Mannock figure was an RAF Reservist and CPGB member, John Loverseed. Returning to Fighter Command, he flew Hurricanes during the fall of France, and was Common Wealth MP for Eddisbury 1943–5.
107. Ben Pimlott, *Hugh Dalton* (London: Jonathan Cape, 1985) pp. 225–67. Bevin's support for Dalton over rearmament was rooted in a similar deep-seated antipathy towards Germans.
108. Ward, *Red Flag and Union Jack*, p. 202.
109. The Labour movement as a whole (i.e. including the trade unions) and not solely the ILP, which was officially opposed to the war. George Orwell, 'My

Country Right or Left' (autumn 1940), *The Collected Essays, Journalism and Letters of George Orwell, Volume 1 An Age Like This 1920–1940* (London: Secker & Warburg, 1968; Penguin, 1970) pp. 589–92.
110. For an argument as to why a 'good war' is particularly associated with Second World War British narratives, see Hynes, *The Soldier's Tale*, pp. 123 and 144–5.
111. Englander and Osborne, 'Jack, Tommy, and Henry Dubb: the Armed Forces and the Working Class', p. 621.
112. C.R. Attlee, *As It Happened* (London: William Heinemann, 1954) p. 97. For a defence of Attlee's handling of defence issues prewar, see Robert Pearce, *Attlee* (London: Longman, 1997) pp. 71–87.
113. Denis Healey, *The Time Of My Life* (London: Michael Joseph, 1989) chapter 13; Susan Crosland, *Tony Crosland* (London: Jonathan Cape, 1982; Coronet, 1983) pp. 13–41, 96, 103, 353 and 382. An example from outside Parliament: E.P. Thompson's intellectual credibility when campaigning for European nuclear disarmament throughout the 1980s was underpinned by his record as a young tank commander four decades earlier.
114. For a forceful argument that Labour maintained its 'traditional contempt of the soldier' throughout the First World War and the whole of the interwar period, see Englander and Osborne, 'Jack, Tommy, and Henry Dubb: the Armed Forces and the Working Class', pp. 593–621, especially pp. 620–1.
115. Kenneth Harris, *Attlee* (London: Weidenfeld & Nicolson, 1982) pp. 34–40, 171–2 and 179–80; Attlee, *As It Happened*, pp. 38–44 and 110–12.
116. Correlli Barnett, *The Audit of War: The Illusion and Reality of Britain as a Great Nation* (London: Macmillan, 1986) pp. 15–18. For a critique of Barnett's belief in an anti-militaristic, anti-technological ' "enlightened" Establishment', see David Edgerton 'The Prophet Militant; The Peculiarities of Correlli Barnett', *Twentieth Century British History*, II (1991) pp. 376–9.
117. Hugh Dalton, *With British Guns in Italy: A Tribute to British Achievement* (London: Methuen, 1919).
118. Ben Pimlott, 'Once upon a lifetime', *Times Higher*, 6 November 1998.
119. Travers, *The Killing Ground*, p. 253.

Bibliography

Contemporary sources

Correspondence of Captain Albert Ball, Nottinghamshire Record Office.
Cornelius Van H. Engert papers, Lauinger Library, Georgetown University.
Frederick Oughton (ed.), *The Personal Diary of 'Mick' Mannock V.C., D.S.O. (2 bars), M.C. (1 bar)* (London: Neville Spearman, 1966).
Chaz Bowyer (ed.), Gwilym H. Lewis, DFC, *Wings Over The Somme 1916–18* (Wrexham: Bridge Books, 1976; 1996 edn).
'Major Robert Gregory MC'; 'Major James McCudden'; 'Major George McElroy'; 'Major Edward Mannock', War Office files on army officers 1914–19, Public Record Office now National Archives.
'1914–1920 RFC/RNAS/RAF decorations', Air Ministry, Public Records Office now National Archives.
Turkey 1914–15, Foreign Office, Public Records Office now National Archives.
Edward Mannock papers, Imperial War Museum.
Independent Labour Party Papers, British Library of Political and Economic Science.

Daily Telegraph
Kentish Gazette
Kentish Observer
Kettering Evening Telegraph
Labour Leader
London Gazette
Observer
Reynold's News
Sunday Graphic and Sunday News
The Times

Almack, Edward, *The History of the Second Dragoons 'Royal Scots Greys'* (London: private publication/public subscription, 1908).
Anon, *War Birds – Diary of an Unknown Aviator* (London: Hamilton, 1926).
Army, *War Office Committee of Enquiry Into 'Shell-Shock'* (London: HMSO, 1922).
Attlee, C.R., *As It Happened* (London: William Heinemann, 1954).
Auden, W.H., *The English Auden: Poems, Essays, & Dramatic Writings, 1927–1939* (London: Faber, 1977).
Bartlett, F.C., *Psychology and the Soldier* (Cambridge: Cambridge University Press, 1927).
Bishop Major William, VC, *Winged Warfare* (London: Hodder & Stoughton, 1918).
Buchan, John, *Mr Standfast* (London: Thomas Nelson, 1919; Penguin, 1997).
Canterbury 1905 (Canterbury: Parker, 1997), a facsimile edition of Canterbury Chamber of Trade, *The Ancient City of Canterbury* (Canterbury: Cross & Jackman, 1905).
Clarke, H. (ed.), 'Rules of the Air and Hints on Flying: Instructions for Students of the CFS Detachment at RFC Netheravon', *Cross and Cockade*, XXIII (1992).

Dalton, Hugh, *With British Guns in Italy: A Tribute to British Achievement* (London: Methuen, 1919).
Graves, Robert, *Goodbye to All That* (London: Jonathan Cape, 1929; Penguin, 1960).
Hillary, Richard, *The Last Enemy* (London: Macmillan, 1942; Pimlico, 1997).
HMSO, *Statistics of the British Empire During the Great War* (London: HMSO, 1922).
Johnson, Air Vice-Marshal 'Johnnie', *Wing Leader* (London: 1956; Goodall, 1995).
Jones, H.A. *The War in the Air Being the Story of the Part Played in the Great War by the Royal Air Force, Volume V History of the Great War Based on Official Documents by Direction of the Historical Section of the Committee of Imperial Defence* (Oxford: Oxford University Press, 1937).
Jones, Ira *An Air Fighter's Scrapbook* (London: Nicholson & Watson, 1938; Greenhill Books, 1990).
Jones, Ira, *King of Air Fighters: The Biography of Major 'Mick' Mannock, VC, DSO, MC* (London: Ivor Nicholson & Watson, 1934; Greenhill Books, 1989).
Jones, Ira, *Tiger Squadron* (London: W.H. Allen/White Lion, 1972).
Kavanagh, P.J. (ed.), *Collected Poems of Ivor Gurney* (Oxford: Oxford University Press, 1982).
Kipling, Rudyard, *Kim* (London: Macmillan, 1901).
Koestler, Arthur, 'In Memory of Richard Hillary', *The Yogi and the Commissar* (London: Jonathan Cape, 1945; paperback edn 1964).
Legge-Pomeroy, Major Ralph, *The Story of a Regiment of Horse: being the Regimental History from 1865 to 1922 of the 5th Princess Charlotte of Wales' Dragoon Guards Volume 1* (William Blackwood & Sons, 1924).
Lewis, Cecil, *Sagittarius Rising* (London: Peter Davies, 1936; Warner, 1994).
McCudden, James Thomas Byford, *Five Years In The Royal Flying Corps* (London: The Aeroplane & General Publishing Co. Ltd, 1918).
MacCurdy, J. T., *War Neuroses* (Cambridge: Cambridge University Press, 1918).
MacDonald, J. Ramsay, *Socialism* (London: ILP, 1907).
'McScotch', *Fighter Pilot* (London: Routledge, 1936; Bath: New Portway, 1972).
National Council for Civil Liberties (Captain Gilbert Hall), *The Story of the Cairo Forces' Parliament* (London: NCCL, 1944).
Orwell, George, 'My Country Right or Left' (autumn 1940), *The Collected Essays, Journalism and Letters of George Orwell, Volume 1 An Age Like This 1920–1940* (London: Secker & Warburg, 1968; Penguin, 1970).
Orwell, George, 'The English Revolution', in *The Lion and the Unicorn: Socialism and the English Genius* (London: Secker & Warburg, 1941; Penguin, 1988).
Rivers, W. H. R., 'Fear and its expression', *Instinct and the Unconscious* (Cambridge: Cambridge University Press, 1922).
de Saint-Exupéry, Antoine, *Flight to Arras* (*Pilote de guerre*, 1942) (London: Penguin, 1995).
Sassoon, Siegfried, *Memoirs of a Fox-hunting Man* (London: Faber & Faber, 1928; paperback edn 1960).
Tawney, R. H., *The Acquisitive Society* (London: G. Bell & Sons Ltd, 1921).
Tressell, Robert, *The Ragged Trousered Philanthropists* (London: Panther, 1965).
War Office, *Alphabetical List of Recipients of the Victoria Cross, during the War, 1914–20* (London: HMSO, 1920).
Wintringham, T. H., 'Mutiny', *Left Review*, August 1935.
Wintringham, Tom, *English Captain* (London: Faber & Faber, 1939).

Yeates, V. M., *Winged Victory* (London: Jonathan Cape, 1934; Southampton: Ashford, Buchan & Enright, 1985).
Yeats, W. B., *Selected Poetry* (London: 1991).

Secondary sources

Banville, John, *The Untouchable* (Picador, 1997).
Barker, Pat, *Regeneration* (London: Viking, 1991; Penguin, 1992).
Barker, Ralph, *The Royal Flying Corps in France: From Bloody April 1917 to Final Victory* (London: Constable, 1995).
Barnett, Correlli, *The Audit of War: The Illusion and Reality of Britain as a Great Nation* (London: Macmillan, 1986).
Bence-Jones, Mark, *The Catholic Families* (London: Constable, 1992).
Blewett, Neal, *The Peers, the Parties and the People: The General Elections of 1910* (London: Macmillan, 1972).
Bogacy, Ted, 'War Neurosis and Cultural Change in England 1914–1922: the work of the War Office Committee of Enquiry Into "Shell-Shock"', *Journal of Contemporary History*, XXIV (1989).
Bourke, Joanna, *Dismembering the Male: Men's Bodies, Britain and the Great War* (London: Reaktion Books, 1996).
Bourke, Joanna, *An Intimate History of Killing: Face-to-face Killing in Twentieth-century Warfare* (London: Granta, 1999).
Bowyer, Chaz, *Albert Ball, VC* (Wrexham: Bridge Books, 1994).
Brereton, J. M., *A History of the $4^{th}/7^{th}$ Royal Dragoon Guards and Their Predecessors* (Catterick: 4th/7th Royal Dragoon Guards, 1982).
Brickhill, Paul, *Reach For The Sky* (London: William Collins, 1956).
Bridgeman, Brian, *The Flyers* (Swindon: Brian Bridgeman, 1989).
Butler, Derek (ed.), *Canterbury In Old Photographs* (Gloucester: Alan Sutton).
Calder, Angus, *The Myth of the Blitz* (London: Pimlico, 1991).
Calder, Angus (ed.), *Wars* (London: Penguin, 1999).
Canadian Air Force Office of Public Affairs, 'The Flying Career of William Avery Bishop', http://raven.cc.ukans.edu/~kansite/ww_one/comment/bishop.html.
Cooksley, Peter G., 'J.T.B. McCudden', *VCs of the First World War: The Air Aces* (Stroud: Sutton Publishing, 1996).
Crang, J. A., 'Army Education and the 1945 General Election', *History*, LXXXI (1996).
Crosland, Susan, *Tony Crosland* (London: Jonathan Cape, 1982; Coronet, 1983).
Davidson, Bill (Sergeant), 'Essay in Labour History: the Cairo Forces Parliament', *Labour History Review*, LV (1990).
Davies, Pete, *Catching Cold: 1918's forgotten tragedy and the scientific hunt for the virus that caused it* (London: Michael Joseph, 1999).
Douglas, Lord, of Kirtleside, *Years of Combat* (London: Collins, 1963).
Dowse, R. E., *Left In The Centre: the Independent Labour Party 1893–1940* (London: Longman, 1966).
Dudgeon, James, '"Mick" Fact, Fiction and Legend: an examination of Major Edward Mannock VC DSO MC', *Cross and Cockade* (1980) XI.
Dudgeon, James M., *'Mick' The Story of Major Edward Mannock V.C., D.S.O., M.C., R.F.C., R.A.F.* (London: Robert Hale, 1981; paperback edn 1993).

Dyer, Geoff, *The Missing of the Somme* (London: Hamish Hamilton, 1994; Penguin, 1995).
Edgerton, David, 'The Prophet Militant; The Peculiarities of Correlli Barnett', *Twentieth Century British History*, II (1991).
Englander, David and Osborne, James, 'Jack, Tommy, and Henry Dubb: the Armed Forces and the Working Class', *Historical Journal*, XXI (1978).
Faulks, Sebastian, *The Fatal Englishman: Three Short Lives* (London: Hutchinson, 1996; Vintage, 1997).
Ferguson, Niall, *The Pity of War* (London: Allen Lane, The Penguin Press, 1998).
Franks, Norman, *Who Downed the Aces in WWI?* (London: Grub Street, 1996).
Hamer, A. C., *A Short History of No. 85 Squadron* (Northampton: Services Publishing Services, 1987).
Harris, B., *The Health of the School Child: a History of the School Medical Service in England and Wales* (Buckingham: Open University Press, 1995).
Harris, Kenneth, *Attlee* (London: Weidenfeld & Nicolson, 1982).
Healey, Denis, *The Time Of My Life* (London: Michael Joseph, 1989).
Heller, Joseph, *British Policy Towards the Ottoman Empire* (London: Frank Cass, 1983).
Holden Reid, Brian, 'T.E. Lawrence and his Biographers', in Brian Bond (ed.), *The First World War and British Military History* (Oxford: Clarendon Press, 1991).
Hopkin, Deian, 'The Labour Party Press', in K.D. Brown (ed.), *The First Labour Party 1906–14* (London: Croom Helm, 1985).
Hynes, Samuel, *The Auden Generation: Literature and Politics in England in the 1930s* (London: Bodley Head, 1976).
Hynes, Samuel, *A War Imagined: The First World War and English Culture* (London: Bodley Head, 1990).
Hynes, Samuel, *The Soldiers' Tale: Bearing Witness to Modern War* (London: Pimlico, 1998)
Joseph, Lyn, *The Roses of No-Man's Land* (London: Michael Joseph, 1980; Papermac, 1984).
Keegan, John, *The Face of Battle: A Study of Agincourt, Waterloo and the Somme* (London: Jonathan Cape, 1976; Pimlico, 1991).
Kelleher, George, D., *The Gunpowder at Ballincollig* (Inniscarra, Co. Cork: Kelleher, 1997).
King, Alex, *Memorials of The Great War in Britain: The Symbolism and Politics of Remembrance* (Oxford: Berg, 1998).
Leed, Eric J., *No Man's Land: Combat and Identity in World War I* (Cambridge: Cambridge University Press, 1979).
Leonard, Jane, 'The Reaction of Irish Officers in the British Army to the Easter Rising of 1916', in Hugh Cecil and Peter H. Liddle (eds), *Facing Armageddon: The First World War Experienced* (London: Leo Cooper, 1996).
Lyle, Marjorie, *Canterbury* (London: Batsford/English Heritage, 1994).
Macfie, A. L., *The End of the Ottoman Empire, 1908–1923* (London: Longman, 1998).
McKibbin, Ross, *The Evolution of the Labour Party* (Oxford: Oxford University Press, 1974).
McKibbin, Ross, 'Why was there no Marxism in Britain?', *The Ideologies of Class: Social Relations in Britain 1880–1950* (Oxford: Oxford University Press, 1990).
Maycock, Clifford, *The Parish Church of St Mary the Virgin Wellingborough*, and *Notes on St Mary the Virgin Wellingborough* 1 and 2, church pamphlets, no date.

Morrow, John H. Jr, 'Knights of the Sky: The Rise of Military Education', in Frans Coetzee and Marilyn Shevin-Coetzee (eds), *Authority, Identity and the Social History of the Great War* (Oxford: Berg, 1995).
Norman, Edward, *Church and Society in England 1770–1970* (Oxford: Oxford University Press, 1976).
Osborne, Charles, *W. H. Auden: The Life of a Poet* (London: Eyre Methuen, 1980).
Oughton, Frederick, and Smyth, Commander Vernon *Ace With One Eye: The Life and Combats of Major Edward Mannock* (London: Frederick Muller, 1963).
Palmer, Joyce and Maurice, *A History of Wellingborough* (Earls Barton: Steepleprint, 1972).
Palmer, Joyce and Maurice, *Wellingborough Memories* (Wellingborough: W.D. Wharton, 1995).
Parker, R.A.C., *Chamberlain and Appeasement: British Policy and the Coming of the Second World War* (London: Macmillan, 1993).
Paris, Michael, 'The Rise of the Airmen: the Origins of Air Force Elitism c. 1890–1918', *Journal of Contemporary History*, XXVIII (1993).
Paris, Michael, *From the Wright Brothers to Top Gun: Aviation, Nationalism and Popular Cinema* (Manchester: Manchester University Press, 1995).
Pearce, Robert, *Attlee* (London: Longman, 1997).
Pelling, Henry, *Social Geography of British Elections 1885–1910* (London: Macmillan, 1967).
Penrose, Harold, *British Aviation: Widening Horizons 1930–1934* (London: HMSO, 1979).
Petter, Martin, ' "Temporary gentlemen" in the aftermath of the Great War: rank, status and the ex-officer problem', *Historical Journal*, XXXVII (1994).
Pierce, David, *Yeats's Worlds: Ireland, England and the Poetic Imagination* (New Haven, CT: Yale University Press, 1995).
Pimlott, Ben, *Hugh Dalton* (London: Jonathan Cape, 1985).
Rang IV, Scoil Mhaire, 'The Church of St Mary and St John', *Times Past*, VI (1989–90).
Reeves, Nicholas, 'The Real Thing At Last: *Battle of the Somme* and the domestic cinema audience in the autumn of 1916', *The Historian*, 51 (1996).
Rhodes, James Robert, *Gallipoli* (London: Batsford, 1965; Pimlico, 1999).
Robinson, Derek, *Goshawk Squadron* (London: Heinemann, 1971; HarperCollins, 1993).
Simpkins, Peter, 'Everyman at War: Recent Interpretations of the Front Line Experience', in Brian Bond (ed.), *The First World War and British Military History* (Oxford: Clarendon Press, 1991).
Smith, Adrian, 'Ramsay MacDonald: Aviator and Actionman', *The Historian*, 28 (1990).
Smith, Adrian, *The New Statesman Portrait of a Political Weekly 1913–1931* (London: Frank Cass, 1996).
Sotheby's, *Historic Aircraft, Aeronautica and Medals*, Sotheby's catalogue, 19 September 1992.
Spiers, Edward, 'The Late Victorian Army' in David Chandler and Ian Beckett (eds), *The Oxford Illustrated History of the British Army* (Oxford: Oxford University Press, 1994).

Spiers, E. M., 'Army Organisation and Society in the Nineteenth Century', in Thomas Bartlett and Keith Jeffery (eds), *A Military History of Ireland* (Cambridge: Cambridge University Press, 1996).

Standage, Tom, *The Victorian Internet: the remarkable story of the telegraph and the nineteenth century's online pioneers* (London: Weidenfeld & Nicolson, 1998).

Nigel Steel, and Hart, Peter, *Tumult In The Clouds: The British Experience of the War in the Air, 1914–1918* (London: Hodder & Stoughton, 1997).

Stokes, D., *Paddy Finucane: Fighter Ace* (London: William Kimber, 1983; Crécy Books, 1992).

Ray, Sturtivant, 'British Flying Training in World War I', *Cross and Cockade*, XXV (1994).

Sturtivant, Ray and Page, Gordon, *The S.E.5 File* (London: Air-Britain, 1996).

Summers, Anne, 'Militarism in Britain', *History Workshop Journal*, 2 (1976).

Sweetman, John, 'The Smuts Report of 1917', *Journal of Strategic Studies*, IV (1981).

Tanner, Duncan, *Political Change and the Labour party, 1900–1918* (Cambridge: Cambridge University Press, 1990).

Thorpe, Andrew, *A History of the Labour Party* (London: Macmillan, 1997).

Travers, Tim, *The Killing Ground: the British Army, the Western Front and the Emergence of Modern Warfare 1900–1918* (London: Allen & Unwin, 1987).

Trevor-Roper, Hugh, 'History and Imagination', in Valerie Pearl, Blair Warden and Hugh Lloyd-Jones (eds), *History and Imagination: Essays in Honour of H.R. Trevor-Roper* (London: Duckworth, 1981).

Ward, Paul, *Red Flag and Union Jack Englishness, Patriotism and the British Left, 1881–1924* (Woodbridge: Boydell Press/Royal Historical Society, 1998).

Wilkinson, Alan, *Christian Socialism: Scott Holland to Tony Blair* (London: SCM Press, 1998).

Williamson, Anne, *Henry Williamson: Tarka and the Last Romantic* (Stroud: Alan Sutton, 1995).

Winstanley, Michael, *Life in Kent At The Turn Of The Century* (Folkestone: Dawson, 1978).

Winter, J. M., *Socialism and the Challenge of War: Ideas and Politics in Britain, 1912–18* (London: Routledge & Kegan Paul, 1974).

Winter, Jay, *Sites of Memory Sites of Mourning: The Great War in European Cultural History* (Cambridge: Cambridge University Press, 1995).

Yeo, Stephen, 'A New Life: The Religion of Socialism in Britain 1883–1896', *History Workshop Journal*, 4 (1977).

Index

Abse, Leo, 52
Aces High (1976), 152–3
aerial combat, mythologizing and romanticizing, 136–7
air aces, 1914–18, 59
Air Organization Committee, 1917, 101
aircraft production, 1914–18, 6
Ali Hamid Bey, 48
Anglo-Catholicism, 33–4
Anglo-Turkish (Ottoman) relations, 1914–15, 40–1, 42–8
Arras memorial, 131
Ashdown, Paddy, 160
Associated-Rediffusion, 152
Asquith, Anthony (*Reach For The Stars*), 151
Asquith, Henry Herbert, 52
Attlee, Clement, 36, 159, 161
Auden, W.H., 147
aviator myth in 1930s, 147–8

Baden Powell, Robert, 14
Bader, Douglas, 4, 16, 143, 144, 151
Baldwin, Stanley, 145
Ball, Albert, 4, 59–60, 63, 67, 73, 74–5, 84, 96, 104, 106, 135, 138, 145, 156
Ballincollig, Co. Cork, 11
Banville, John (*The Untouchable*), 8
Barker, Pat, 1
Barnett, Correlli, 160
Bartlett, F.C., 121, 122
Battle of Britain, 1940, 143
The Battle of the Somme (1916), 50
Belloc, Hilaire, 34
Bennett, Arnold, 30, 39
Bennett-Goldney, Francis, 22–3
Bertrab, Joachim von, 86, 89
Bevin, Ernest, 158, 161
'Big Wing' strategy, 1940, 143
Bishop, William 'Billy', 4, 59, 67, 73, 75, 96, 104, 106, 115, 116, 119, 122, 123, 135, 136, 137–8

Blaxland, Lionel, 76
'Bloody April', 1917, 5, 60, 66–7, 84
Boelcke, Oswald, 60, 68, 144
Boer War, 14–15, 22
British Union of Fascists (BUF), 157–8
Brockway, Fenner, 28
Buchan, John (*Mr Standfast*), 56, 136, 144
Buchanan, J.E., 55, 56–7

Cairo Forces Parliament, 1944, 51–2
Cairns, W.E., 116
Caldwell, Keith 'Grids', 17, 106–7, 109, 110, 111–12, 113, 114, 115, 116, 118, 123, 128, 132, 133–4, 151, 155
Callaghan, James ('Jim'), 160
Campbell, Stewart, 53
Canadian Army Corps, 67, 88, 90
Canterbury
  Cavalry Depot, 15
  commemoration of Mannock, 5, 134–5, 139–42, 142, 143, 162
  Edwardian cathedral city, 18–19
  Edwardian municipal politics, 22–4
  Cathedral memorial service, 141–2, 162
  Cathedral memorial tablet, 141, 162
Chamberlain, Neville, 147, 158
Churchill, Winston, 134–5
cinema representation of aerial warfare, 136, 148, 152–3
Clare, John, 132
class composition of British Army officers, 1914–18, 54
Clements, H.G., 17, 107, 113, 125, 133
commemoration and remembrance, 5
Comper, Sir John Ninian, 32–3
Co-operative Union ('Co-op'), 27
Conservative Party and matters military, 160
Constantinople, 40, 66, 81
Craig, E.W., 121, 122

205

Craiglockhart War Hospital, 120
Cripps, Sir Stafford, 161
Crosland, Anthony 'Tony', 160
Cross and Cockade International, the First World War Aviation Society, 153
Crowe, Eyre, 45
Cunningham, John 'Cats Eyes', 4

Daily Herald, 28, 79
Daily Mail, 50
Dalton, Hugh, 158–9, 161
Dolan, 109, 112, 114, 115, 124, 137
Dore, Alan, 106
Douglas, Sholto, 114, 150
Douglas, William, 90
Douglas-Watson, F., 41, 53
Dudgeon, James M., 2, 4, 15, 47–8, 119, 125, 126, 129–30, 133, 137, 151, 153, 155

Easter Rising, Dublin, 1916, 55–6
East Kent Yeomanry, 22
Engert, Cornelius Van Hemert, 46–7
Enver Pasha, 41, 43
Eyles, A.E. ('Jim')
  accompanying Mannock to London, January 1918, 101
  collaboration with Ira Jones, 103
  background, 28
  education of Mannock, 30
  enquiries re Mannock's grave, 130–1
  farewell to Mannock, February 1914, 40
  final leave of Mannock, June 1918, 119–20
  first leave of Mannock, summer 1917, 79
  leave early 1918, 100
  petitioning for VC for Mannock, 1919, 139, 154, 157, 159
  relative (Mrs E. Woolford), 152
  shared home with Mannock, 28–30
  socialism, 37–8, 107
Eyles, Derek, 28, 29, 30
Eyles, Mabel, 28, 29, 107, 139, 159

Fabian Society, 27, 30, 35

Falklands War, 1982, 160
Faulks, Sebastian, 1
Finucane, Brendan, 'Paddy', 144
Foncke, René, 138
Foot, Michael, 160
Frech, Fritz, 89
French, Field Marshal Sir John, 52

Gaitskell, Hugh, 160
Galsworthy, John, 30
Gallipoli/Dardanelles campaign, 1915, 45, 47
Gardner, Cuthbert, 20–1, 22, 24, 129, 134, 140
German offensive, March 1918, 108–9, 114–15, 117
German technological superiority 1916–17, 68–9, 73
Gibson, Guy, 4
Glasier, Bruce, 27
Godfrey, Steve, 82, 88–9
Green, T.H., 81
Gregory, Lady, 98
Gregory, Robert, 97–8
Guilty Men, 160

Haig, Field Marshal Sir Douglas, 58, 92, 109
Haldane, Richard B., 22
Hall, Gilbert, and 1944 Cairo Forces Parliament, 51–2
Halton Park Camp West, Berkshire, 1914–15, 49–53
Hardie, Keir, 27, 30, 36, 55
Hawkins, Jack, 151
Headlam, Stewart, 35
Healey, Denis, 160
Heaton, John Henicker, 22–3
Henchley, A.R., 22
Henderson, Arthur, 37, 71, 159
Henderson, Major-General Sir David, 101–2, 122, 143
Hews, G.R., 140
Hillary, Richard (*The Last Enemy*), 4, 16, 148–50
Horn, K.K. 'Nigger', 116, 157
Howe, 'Swazi', 109–110, 157
Hyndman, H.M., 36
Hynes, Samuel, 147, 148

Imperial War Graves Commission
  (IWGC), 129–31, 133
Independent Labour Party (ILP), 27–8,
  31–2, 157, 161
  Canterbury/East Kent ILP, 23–4
  Glasgow Shop Stewards Movement,
    38
  Wellingborough ILP, 27–8, 31–2,
    37–8
  see also Labour Party
India, 14–16
Inglis, Donald, 126–8, 129, 132
International Brigade, 158, 159

J. Arthur Rank, 152
Johnson, Amy, 147
Johns, Captain W.E. (*Biggles* novels),
  16, 146
Johnson, Dean Hewlett, 141
Johnson, 'Johnnie', 139, 143
Jones, Ira 'Taffy'
  as a biographer, 2, 4, 8, 17, 47–8, 65,
    71, 103–4, 113, 116, 118–19,
    123, 124, 125, 129, 133, 137–8,
    143, 144–7, 151, 154–6, 157
  as a fellow pilot of Mannock in 74
    Squadron, 103–5, 111, 113, 118,
    125, 126, 128

Kent and Canterbury Hospital, 140–1
Kent Tariff Reform League, 23
Keymer, Padre Bernard W., 76, 128
Kinnock, Neil, 160
Kipling, Rudyard (*Kim*), 15
Kirkwood, David, 50

Labour Leader, 28
Labour Party, 24, 27–8, 37–8, 139
  and military matters, 9, 36–7, 158–62
  see also Independent Labour Party
    (ILP)
Lansbury, George, 28, 157, 159
Laventie military cemetery, 130, 132,
  133
Lawrence, T.E. 'Lawrence of Arabia',
  135–6, 147, 148, 149, 151
Le Queux, William, 50
Lean, David (director, *The Bridge Over
  the River Kwai*), 47

Ledwidge, Francis, 55
Leefe Robinson, William, 135
Lewis, Cecil (*Sagittarius Rising*), 2, 81,
  146, 148, 153
Lewis, Gwilym, 61, 91–2, 93, 94,
  95, 98–9, 112, 115–16, 117,
  122, 125
Lewis, Hugh, 117
Lewis, Mary, 117, 156, 157
Liberal Democrats, 160
Lloyd, George 'Zulu', 76–7, 84–5, 87,
  92, 113
Lloyd George, David, 30, 37, 136
Ludendorff, General Erich von
Luftwaffe, warning of 1930s expansion
  and threat, 144–5
Lutyens, Edwin, 138, 140

Malan, Adolph 'Sailor', 144
MacCudden, James, 4, 6, 7–8, 59, 63,
  65, 89–90, 96–7, 106, 115, 116–17,
  118, 121, 122, 127, 135, 138, 143,
  148, 149, 156
MacCurdy, J.T., 121, 122
MacDonald, Ramsay, 9, 27–8, 30, 35,
  36, 71, 159
McElroy, George 'McIrish', 82, 90–2,
  97, 125, 137
McGregor, Malcolm, 128
McKenna, Reginald, 134–5
McKenna, Virginia, 151
MacLanachan, William 'McScotch'
  as a biographer, 2, 4, 17, 63, 84, 86,
    116, 133–4, 138, 143, 144–5,
    146, 154–6
  as fellow pilot of Mannock in 40
    Squadron, 80, 82, 83–4, 86–90,
    92, 95, 97, 98–9, 100, 156
McLaughlin, Edgar, 117
Macmillan, Norman, 114
Mallett, Sir Louis, 41, 43, 44
Mannock, Dorothy, 120
Mannock/Corringham, Edward, 11,
  12–13, 14–15, 138
Mannock, Edward 'Mick'/ 'Pat'/
  'Paddy'/ 'Jerry'/ 'Murphy'
  'ace with one eye' myth, 15–18, 32,
    57, 78, 133–4
  'anxiety neurosis', 112–13, 118–22

Mannock, Edward (*continued*)
  arrival on the Western Front, April 1917, 66–7
  biographies of, 2, 3, 143, 144–5, 146–7, 150–1, 153–4
  birthplace confusion, 13–14
  commemoration in Canterbury, 139–42, 143, 162
  commemoration in Wellingborough, 5, 138–9, 142
  comparison with Albert Ball, 60, 63, 73, 74–5, 105, 114, 115, 135, 138
  comparison with 'Paddy' Finucane, 144
  comparison and rivalry with James McCudden, 7–8, 63, 89–90, 96–7, 106, 115, 116–17, 118, 121, 127, 135, 138
  death, 26 July 1918, 126–8, 129–30, 132–3
  eagerness in early 1918 to return to active service, 100–2
  employment in Canterbury, 19–20
  employment in Wellingborough, 25–6
  Eyles 'second family' 28–30
  farewell to 40 Squadron, January 1918, 98–9
  fear of a 'flamer' (burning to death), 72, 89–90, 92, 111, 112–13, 120–1, 133
  fictional treatment (including film), 5, 151–4
  flight commander, 40 Squadron, 85–93
  flight commander, 74 Squadron, 106–18
  flying ability, 63
  Germans loathed, 48, 112, 117, 154–6
  grave, disputed, 129–31, 132, 133
  Halton Park mock parliament, 1915, 51–2
  Indian childhood, 15, 18
  inexperience and fear, spring 1917, 71–4, 84–5
  Irish identity and origins, 14, 30, 55–6, 91, 155
  Keymer, Padre, friendship, 76, 154

  'kills', final tally, 134, 136–7
  Lewis, Mary, friendship and correspondence, 117, 156
  Lloyd, George 'Zulu', friendship, 76–7, 82, 84–5, 87, 92, 113
  McCudden's advice on how to spin acted on, 65–6
  marksmanship and gunnery practice, 73–4, 87, 115
  military honours, 82, 90, 115, 119, 134–5, 136–7
  mythologizing of, 4, 47–8, 134, 136–7, 143, 145, 146–7
  musicianship, 30
  political beliefs (socialism), 5, 18, 24, 30–2, 35–6, 37, 50–3, 54–5, 80–2, 113–14, 154–6, 161–2
  posting to 74 Squadron, January 1918, 104
  posting to 85 Squadron, June 1918, 119, 122
  problems researching, 3–4
  reasons for writing about, 9–10, 16–17
  relationship with his father, 18
  religious upbringing and beliefs, 15, 19, 24, 32–6
  repatriation from Constantinople, 1915, 46
  Royal Engineers Signal Section service, 1916, 53–7
  RFC/RAF Club social activities, 79
  sale of medals and artefacts, 138, 139
  'Society', antipathy towards, 80–1
  squadron commander, 85 Squadron, June–July 1918, 123–8
  speculation on postwar career, 8, 156–8
  summary of wartime career and awards, 2–3
  survival on the Western Front, 6
  tactical innovator and influence on Fighter Command, 5, 75, 84–5, 87, 90, 92, 104–6, 110–11, 113–14, 123–4, 143–4, 149, 153
  Territorial Force service (2nd Home Counties Field Ambulance, RAMC), 21–2, 31, 48–53
  training to fly, 61–6

training novice pilots, 1918, 104–6, 112, 113
transfer from RAMC to Royal Engineers and commissioned, winter 1915–16, 53–4
transfer from Royal Engineers to Royal Flying Corps, 1916, 57–60
trenches, impressions of, 83, 90
Trenchard, antipathy towards, 79, 94–5
Turkish employment, 1914, 39–41, 44
Turkish social life, 42
Turkish internment, 44–8, 154
unpopularity initially in the RFC, 63, 70–1
Wellingborough ILP involvement, 31–2, 37–8, 107
women friends, 42, 84, 117, 125
working-class background, 7–8
Mannock, Frank, 56
Mannock, Jess, 13, 78, 100, 118
Mannock, John P., 12, 80
Mannock, Julia (Sullivan), 11, 12–13, 14, 15, 19, 49, 76, 78, 100, 128
Mannock, Nora, 19, 78
Mannock, Patrick [brother], 13, 20, 40, 78, 120, 128, 138, 141
Mannock, Patrick [cousin], 79–80
Maxton, James, 157
Maxwell, W.H., 121
Menin Gate memorial, 131
Mills, John, 151
militarism and socialism before 1914, 36–7
Minter, Florence, 47
Money, Leo Chiozza, 37–8
Morgenthau, Henry, 44–5, 46
Morris, William, 34–5
Mosley, Sir Oswald, 157

National Government, 1931–40, 145
National Service League, 22
National Telephone Company, 20, 22, 25, 40–1, 43, 56
Naulls, Edward, 129
'New Liberalism', 81
Nicolson, Harold, 42
Nieuport 17 scout aircraft, 97

comparison with German aircraft, 67, 69, 73

'Operation Michael', German offensive, March 1918, 108–9, 114–15
Orwell, George, 150, 159
Oughton, Frederick, 2, 4, 15, 47–8, 51, 58, 65, 126, 150–1, 152

Paris Michael, 136
Pearce, J.L.T., 94
pilots, RFC/RAF
education, 2
life expectancy, 5–6, 69–70, 108–9
training, 58, 61
Pimlott, Ben, 161
Plumer, General Sir Herbert, 115
Pope Leo XIII, author of 1891 *Rerum Novarum*, 34
Powell, Michael (director, *A Canterbury Tale*), 151
Pressburger, Emeric (writer, *A Canterbury Tale*), 151

Randall, A.C., 128
Rattigan, Terence (*Ross*), 151
Rawson, Albert, 21
Rawson, Fred, 21
'Red Clydesiders' (ILP), 157
Red Cross, 131
Reynolds New, 79
Rhys Davids, Arthur, 80, 156
Richthofen, Lothar von, 60
Richthofen, Baron Manfred von, 'the Red Baron', 59–60, 68, 86, 106, 112, 114, 116, 117, 144–5
Rivers, W.H.R., 120–1
Roberts, Field Marshal Lord, 36
Roberts, Peter (*The Action of the Tiger*), 152
Robinson, Peter (*Goshawk Squadron*), 153
Rothermere, Lord, 102, 136
Royal Air Force (RAF)
creation, 1 April 1918, 108
expansion in the 1930s, 144–5
Fighter Command in the Second World War, 143–4, 150

Royal Air Force Association
  (Canterbury branch), 141–2
Royal Flying Corps (RFC)/Royal Air
  Force (RAF)
  as a meritocracy, 81
  losses, spring 1918, 108–9
  merger of the RFC with the Royal
    Naval Air Service (RNAS), 1 April
    1918, 108
  40 Squadron, 66–7
  56 Squadron, 56–7, 96–7
  74 Squadron, 102–3, 107–8, 110–11,
    123, 125
  see also pilots, RFC/RAF
RFC/RAF Club, 179, 119
Ruskin, John, 34

Sackville-West, Vita, 42
Saint-Exupéry, Antoine de (*Flight To
  Arras*), 148, 149
Sakai, Saburo, 133
Sassoon, Siegfried 'Mad Jack', 121
Saunders, Liman von, 41
SE5a (and SE5) scout aircraft, 6–7, 60,
  93–5, 96, 97
Shaw, George Bernard, 30, 55
Shephard, Gordon, 95
Sheriff, R.C. (*Journey's End*), 152, 153
Sister Flanagan, 125, 151
Smyth, Vernon, 2, 4, 8, 15, 47–8, 58,
  65, 119, 125, 126, 150–1
Smuts, General Jan Christian, 101
Snowden, Philip, 23, 27, 157
Social Democratic Federation (SDF),
  36, 37
Socialism, Christian, 34–5, 37
Société Anonyme Ottoman des Téléphones
  (Constantinople Telephone
  Company), 39–40, 41–2, 43
Sopwith Camel scout aircraft, 7, 68, 69,
  97, 110
Spanish Civil War, 157, 159, 160
Spanish influenza, 122–3
Speed, Walter, 23–4
St Mary the Virgin CofE church,
  Wellingborough, 32–5, 138
Steel, Nigel, and Hart, Peter (*Tumult In
  The Clouds*), 1, 7–8, 97
Stewart, Angus, 132

Stockholm Conference, 71
Swart, Major, 64, 65–6
Sykes, Major-General Frederick H., 102,
  143

tactics in aerial warfare, 60, 67, 69,
  73–4, 84–5, 124–5
Tawney, R.H., 34, 36
Taylor, A.J.P., 150
Thomas, J.H. 'Jimmy', 157
Tilney, Leonard, 67, 70, 75, 76, 82–3,
  85, 89, 93, 95, 98, 100
Todd, Richard, 151
Tompkins, Eric, 58–9
Townsend, Peter, 153
training to fly, 61
Trenchard, Major-General Sir Hugh
  'Boom', 59, 60, 67, 79, 94–5, 96,
  101–2, 111, 136, 143, 145
Tudhop, J.H., 85, 87
Turkish (Ottoman)–British relations,
  1914–15, 40–1, 42–8
Turkish (Ottoman) internment of
  Allied civilians, 1914–15, 44–8

Udet, Ernst, 114–15

van Ryneveld, Colonel Pierre, 110,
  111
Vaughan, Cardinal, 34
Victoria Cross (VC), RAF policy on
  award of, 135
Vimy Ridge, 67, 93
Voss, Werner, 86

War Office Committee of Enquiry on
  'Shell-Shock', 1920–22, 120
Waugh, Evelyn (*Brideshead Revisited*),
  150
Webb, Sidney and Beatrice, 30, 35
Wells, H.G., 30, 39, 55
Wellingborough
  commemoration of Mannock, 5,
    138–9, 142
  Edwardian community and politics,
    26–8, 37–8
'Wellingborough Parliament', 31
Western Front, description and history
  of, 1–2

Western Front Association, 153
Wheatley, John, 34, 50, 157
Whitehead Aircraft Co., 79
Williamson, Henry, 157–8
Wintringham, Tom, 158
Woolford, Eleanor, 152
Workers and Soldiers Councils in the British Army, 1917, 52

Workers Educational Association (WEA), 141
World Disarmament Conference, 145
Wyndham Lewis, Percy, 158

Yeats, W.B., 55, 98, 147
Yeates, Victor (*Winged Victory*), 143–4, 157–8